Resounding Praise for

radical

THE SCIENCE, CULTURE, AND HISTORY OF BREAST CANCER IN AMERICA

"Fabulous...Pickert examines the way we approach screening, prevention, and treatment in America, including the 'awareness' and pinkwashing of breast cancer, and the societal aspects of the disease."

—Jaime Herndon, *Book Riot*

"Riveting...A remarkable, up-to-the-minute resource...An unflinching and clear-eyed examination of scientific and cultural progress and failure."

—Ruth Pennebaker, *Washington Post*

"Want to learn more about breast cancer? Read this book... Pickert has produced an evenhanded, powerful, and unflinching page-turner."

—Pauline Chen, *New York Times Book Review*

"Kate Pickert does a great job of using her own story to bring the people and history of breast cancer treatment and advocacy to a modern audience."

—Susan M. Love, MD, author of *Dr. Susan Love's Breast Book*

"A compassionate, lucid, and well-researched account of historical and ongoing attempts to combat breast cancer."

—*Publishers Weekly*

"A thoroughly researched and compassionate account of a disease that strikes one in eight American women."

—Megan O'Neill Melle, *Parade*

"A must-read if you or anyone you know has breast cancer, but also for anyone interested in the less often discussed factors that influence our perception and care for cancer today."

—Liz Moody, *mindbodygreen*

"Balanced, cogent, and eye-opening." —Barbara Kiser, *Nature*

radical

THE SCIENCE, CULTURE, AND HISTORY OF BREAST CANCER IN AMERICA

kate pickert

Little, Brown Spark
New York Boston London

Little, Brown Spark
Hachette Book Group
1290 Avenue of the Americas
New York, NY 10104
littlebrownspark.com

Originally published in hardcover by Little, Brown Spark, October 2019
First Little, Brown Spark paperback edition, September 2020

Little, Brown Spark is an imprint of Little, Brown and Company, a division of Hachette Book Group, Inc. The Little, Brown Spark name and logo are trademarks of Hachette Book Group, Inc.

10 9 8 7 6 5 4 3 2 1

ISBN 978-0-316-47032-2 (hc) / 978-0-316-47034-6 (pb)
LCCN 2019937030

LSC-C

Printed in the United States of America

To Evangeline

contents

radical

introduction

Of all the things to look at while poison is being pumped into your body, the Pacific Ocean is a pretty good option. The first day I walked into a sixth-floor chemotherapy infusion center in Santa Monica, California, I claimed a pleather recliner in front of a tall window facing due west. I could see Catalina Island just to the south, its sloping hills rising up from the water.

For more than a month, I had been mired in the vicious limbo familiar to most newly diagnosed cancer patients. I was crippled with anxiety about the malignant tumors growing inside me and filled with dread over the assault on my body that would soon begin. It's a strange thing to wait for something terrible to happen when the mechanics of it are entirely unfamiliar. Somehow, I had never known a breast cancer patient. There were women in the outer reaches of my social and professional orbits who had suffered from or died of the disease. But I had never watched it up close. I was in an inhospitable country I did not know and had never visited. I had crossed over into what Susan Sontag called "the kingdom of the sick."

"Plug me in and turn it up," I told a nurse as she screwed a long line of clear tubing into the catheter jutting out from beneath my upper right arm. She smiled and checked Facebook on her iPhone in between changing the bags of liquid that ran from an IV pole through the tubing and into my body. The nurse also used her phone to track how long it took for the liquid to run in, a bell-tower ring marking the end of one drug and the beginning of another. The normalcy of the sound reminded me that, while I was undergoing probably the most important event in my life, I was not particularly special and neither was my disease. Aside from the fact that I was relatively young, I was an unremarkable patient with a common disease, one of about three hundred thousand American women diagnosed with breast cancer in 2014.

While breast cancer is common, its influence on American culture and the scientific community is not. There are symbols and awareness campaigns for scores of causes, but none so recognizable as the pink ribbon in October. Celebrities get diseases all the time, but most Americans would struggle to recite the names of famous people who have suffered from them. Yet for every generation, there is a list of well-known breast cancer patients who told the world about their disease: Betty Ford, Nancy Reagan, Olivia Newton-John, Nina Simone, Linda McCartney, Sheryl Crow, Melissa Etheridge, Elizabeth Edwards, Robin Roberts, Julia Louis-Dreyfus.

In addition to being the most public of all the cancers, carcinoma of the breast may be the most thoroughly studied malignancy in human history. The U.S. National Institutes of Health spends more on breast cancer research than on any other cancer type. The American Cancer Society awards more grants to study breast cancer than any other type of the disease. Along with private spending, funds earmarked for breast cancer research total more than one billion dollars every year. This flood of money has made breast cancer a proving ground for advanced scientific research into the biology and genetics of all cancers and the methods devised to eradicate them. As I type this sentence, 1,823 federally registered breast cancer clinical trials are actively recruiting patients.

On the whole, enduring treatment for breast cancer is easier today than it has ever been, with effective smaller surgeries, less-toxic treatments and reliable drugs that blunt side effects and pain. The breast cancer mortality rate is 35 percent lower today than it was in 1990, and it's dropping steadily, by about 1 to 2 percent per year. We are curing more patients than ever.

And yet, about one-third of all women diagnosed with early-stage invasive breast cancer will eventually see their disease recur or become metastatic. Breast cancer still kills some forty thousand American women every year.

This sobering statistic is a testament to the cleverness of the disease and its ever-growing list of known variations. The malady we call breast cancer is, in reality, a broad category of diseases, each with its own risk profile and sensitivity to drugs. This relatively recent revelation has birthed an oncological race to tailor treatments (and, increasingly, diagnoses) to groups and individuals and apply treatments with greater and greater precision. But scientific progress in breast cancer has been, by any measure, frustratingly slow and incremental. This is not unique to breast cancer. Rather than a straight line toward better and better care, cancer science in general is more like a road up a mountain that's full of switchbacks and blind corners. The story of breast cancer in America is a story of great successes but also many wrong turns. Things work in the lab and fail in patients. Assumptions about the disease that scientists once held as unassailable facts seem ludicrous in retrospect.

Our inability to end the scourge of breast cancer, though, isn't just about the disease's complexity or the slow pace of science. It's also rooted in where our focus has been for the past half century. It's now clear that, in many senses, we have been having the wrong conversations about breast cancer. Screening mammograms are helpful to some women, but they fail to detect some lethal breast cancers; sometimes they catch cancers that are not life-threatening but are treated anyway. Breast cancer awareness is at an all-time high. But despite the attention, there has been no successful large-scale effort to learn what causes breast cancer beyond the small percentage of cases tied to hereditary genetic mutations and the identification of risk factors so ubiquitous among American women that they are virtually useless as warnings. Doctors do not know how or precisely why breast cancer spreads, nor do they know how to kill it off once it does. We have spent relatively few research dollars trying to answer these questions. Despite the races, ribbons and scientific breakthroughs, women are just as confused and frightened as ever.

It's also now clear that in addition to failing to cure many breast cancer patients, treatments women relied on and suffered through for generations were needlessly harsh. After decades of aggressive surgery, chemotherapy and radiation, an approach known as *de-escalation* is now in vogue, with doctors and scientists studying whether it might be safe for certain women to skip some of these common treatments on the way to good health. Mastectomies, which surgeons once believed could remove the threat of breast cancer at its source, are becoming a last resort for patients. It is now possible to imagine a future in which women with the most aggressive forms of breast cancer might be able to safely avoid surgery altogether. Pharmaceutical companies that make breast cancer drugs are largely focused on a single goal—to replace conventional chemotherapy, a treatment that has been the gold standard for more than seventy years, with targeted and gentler drugs.

As a breast cancer patient in 2014 and 2015, I had the old and the new. In addition to chemotherapy, I had a double mastectomy, surgical removal of twenty-two lymph nodes under my arm, five weeks of daily radiation treatment and a full year of targeted intravenous drug therapy. There was little talk of doing less. At thirty-five and with no family history of breast cancer, the odds that I would be diagnosed with an aggressive form of the disease that had already spread to my lymph nodes were so long that they barely registered on any chart. Even my doctors were scared. I was shocked to learn the news and terrified of what it could mean, of course. But during my 372 days as a breast cancer patient, I realized that while I had been extremely unlucky to get a disease that required a sickening, painful and disfiguring course of treatment, I was also clearly very fortunate. If I had been diagnosed with the same disease a decade earlier, before a groundbreaking breast cancer drug became available to the market for patients like me, it's very likely I would have died within a few years. If I had been diagnosed even a year earlier, when my breast cancer was likely present but still undetected, I would have had a different, and possibly less effective, course of treatment. I was lucky in other ways too. I happened to live in Los Angeles, which, in addition to being a world capital of breast-augmentation surgery, is ground zero for research into the type of breast cancer that struck me so unexpectedly.

Less than halfway through my treatment, a pathology report transformed my prognosis from grim to excellent. Drugs and techniques allowed me to avoid side effects that were once hallmarks of breast cancer treatment. I worked throughout my chemotherapy and finished with a full head of hair. A few months after my chemotherapy and surgery and just three days after my final radiation treatment, I started a new job with an employer who had no idea I was a cancer patient.

Despite the familiar archetype of a breast cancer patient who battles her disease, I never felt like I was at war. In fact, there seemed to be very little I could do to participate in my own care. I merely submitted to a largely effective regimen that was the result of decades of trial and error.

Still, I had mental whiplash. My own experience had been, in almost all senses, diametrically opposed to the brand of breast cancer I had learned about through popular media's portrayal of the experience. Before I was diagnosed, in addition to watching tragic breast cancer stories depicted in movies and on television shows, I had even written about the scourge as a reporter for *Time* magazine, parsing new guidelines about mammography screening and the impact Angelina Jolie's double mastectomy might have on other women. But I came to realize that the disease I thought I knew was a mirage. Even though my cancer was aggressive and already spreading when I met the oncologist who directed my care, she said my type was her "favorite" because it was so treatable. At the same time, I learned that breast cancer is a more formidable foe than the races, ribbons and culture around it would have us believe.

This is a book about those contradictions—the triumphs and defeats. It's also about the power of women and what it's like to benefit and suffer as the result of being a woman. It's an attempt to explain how we have understood and tried to tame breast cancer throughout history and about what that history, with its twists and turns and misconceptions, says about our society and a disease inextricably tied to the concept of motherhood. This is also a book about America's particular brand of medicine, bigger and more expensive than care available anywhere else in the world, for better and for worse.

When I told Larry Norton, Memorial Sloan Kettering's top breast cancer doctor, that I was writing a cultural and scientific history of breast cancer in America, he said, "That's not a book. That's an eight-volume set." Fair enough. This is not an encyclopedia of breast cancer, but through the people, places and science presented here, I hope to deepen understanding of a disease so common that to know something about it is to know something about humanity itself.

No names or personal details of anyone depicted in this book have been altered. There are no composite characters, and no timelines have been tweaked for narrative purposes. Quotes from doctor visits and interviews were recorded on tape or, in a handful of instances, documented in contemporaneous notes. To the best of my abilities as a reporter, this is a true story about a disease that touches nearly everyone in America, the doctors and researchers who work against it, and the women caught in the middle.

1

seek and ye shall find

It got to the point where Diana Petitti had to turn her cell phone off just to get some peace. Reporters had been calling every hour, it seemed, and each one had an urgent question that couldn't wait. Thanksgiving had been a bust, ruined by the news that Petitti would soon be grilled by a congressional committee. *Larry King Live* wanted to do a segment and no one could see how that would turn out well.

Petitti had known her work would draw a lot of attention, some of it negative. But death threats? Who could have expected those? *You should die,* said one e-mail message. *I hope you die,* said another. The threats rattled Petitti, but she wasn't scared. Not exactly. She was, more than anything, confused. "It was totally bewildering to me," she said. "Why all the emotion?"

Petitti isn't a person prone to emotion. She's a math geek. She loves the way numbers ground issues, the way they bring clarity to chaos. A few years back, some homes in Petitti's gated Arizona community were burglarized. Residents were terrified. "People were saying we were in the midst of a crime epidemic," said Petitti. But what did the numbers say? She went to look. Petitti downloaded crime statistics from the previous three years, plotted them on graphs and sent the data to the community's homeowners' association. There was no crime wave. Petitti couldn't have been more satisfied. "Numbers are comforting," she said, "because they're real."

Petitti earned good grades and got into Harvard Medical School. But after she finished her residency in internal medicine, she realized the work didn't suit her. "I had more of a fascination with the science of medicine than the aspect where you actually take care of people," she said. A mentor recommended Petitti take some time away from the bedside and join the Epidemic Intelligence Service at the Centers for Disease Control, a program whose officers collect data and investigate disease patterns in order to shape public health policy. Petitti had found her calling. "I just loved epidemiology," she said. "Everything about disease causation, data and information." Epidemiology is the study of how and why diseases occur in populations. Rather than looking through microscopes, epidemiologists analyze statistical patterns and trends for answers.

A few years after her two-year stint at the CDC, Petitti took a job at the University of California, San Francisco, where she taught epidemiology and conducted research. It was the 1980s and the crack epidemic was ravaging many American cities, including Oakland, just across San Francisco Bay. Petitti's work helped demonstrate a link between Oakland's growing crack problem and the troubling increase in low-birth-weight babies born in the county. The numbers brought the drug problem into focus and helped local leaders fight for more funds to combat it. It was thrilling. "Just sitting around and analyzing data was not what I wanted to do," said Petitti. "I wanted to be closer to the ability to take data and turn it into action."

In 1993, Petitti walked away from her tenured university job to work for Kaiser Permanente, a health system known for designing protocols and health interventions around cold, hard numbers. "Well before people saw the value of data for the sake of data, they were linking data and making it part of their decision-making," Petitti said. At Kaiser, Petitti published studies on everything from diabetes to contraception, leveraging data gathered from Kaiser's large patient population to help set policies for doctors and administrators. She even put her obsession with numbers and data to work in long form, publishing a book called *Meta-Analysis, Decision Analysis, and Cost-Effectiveness Analysis: Methods for Quantitative Synthesis in Medicine* in 1994.

The book, which was reissued in 1999, made a splash in the small world of epidemiology, and in 2004 Petitti landed a coveted spot on the U.S. Preventive Services Task Force. The government-funded body evaluates the effectiveness of tests and preventive treatments and makes recommendations to improve health and save lives. The recommendations are not binding, but insurers often use them to set coverage policies. There was no compensation to sit on the task force and the workload was crushing, but Petitti wanted in. The task force was prestigious, powerful, and, until 2009, the year Petitti started getting death threats, almost universally respected by medical professionals and the public.

The idea of looking for breast cancer before it could be seen or felt started with the cervix. In the late 1920s, physician George Papanicolaou found that by scraping a few cells from the opening of a woman's uterus and examining them under a microscope, he could detect early signs of cervical cancer. Back then, the disease was particularly deadly, usually found only after it had spread. Papanicolaou's test, known as a Pap smear, allowed doctors to find and treat cervical cancer early and caused the U.S. death rate from the disease to eventually fall by nearly two-thirds.

If Papanicolaou's strategy of looking for rogue cells before they grew into threatening tumors worked for cervical cancer, why not breast cancer, which killed far more women? Out of fear or ignorance, until the twentieth century, women often did not seek treatment for breast cancer until their tumors had grown so large they were visible to the naked eye or, worse, had broken through the skin to form ulcerating wounds. By that point, the disease, in most cases, had also spread to vital organs and become fatal. But by the 1950s, early detection of breast cancer had become a public-health mantra. Various organizations urged women to monitor their breasts for signs of cancer, with slogans like "Delay kills!" Pamphlets produced by the American Cancer Society advised women to perform self-exams and tell their doctors if they detected lumps or other abnormalities. Some literature included illustrations of tombstones. The message was clear: Look for early breast cancer and live. Ignore the threat and die. Unfortunately, it wasn't that simple. In postwar America, doctors were surgically removing breast tumors earlier than they had in the past but seemingly still not early enough. Women were dying of breast cancer at nearly the same rate as before the public-awareness campaign spread across the country.

Technology offered a possible solution. X-ray machines,

invented in the 1890s, were widely used to confirm diagnoses of breast cancer made through touch. Maybe, just maybe, they could be used to find breast cancers even before they could be felt.

A New York–based health-insurance company launched America's first large-scale breast cancer–screening trial in 1963 using a type of x-ray technology known as mammography. (One of the trial's leaders was a doctor named Philip Strax, whose wife had died of breast cancer in her thirties.) Headquartered in New York City, the program used mammography to provide breast cancer screening to some thirty thousand women aged forty to sixty-four, comparing outcomes to a similar-size group that had not been screened.

In 1971, trial organizers published a paper in the *Journal of the American Medical Association* that appeared to support the theory that screening mammography could save lives. Between 1963 and 1969, thirty-one women in the screening-trial group had died of breast cancer; in the control group, there were fifty-two breast cancer deaths. In reporting the data, the New York researchers wrote that more follow-up was needed to produce definitive conclusions but that the early results were "encouraging" and grounds for "cautious optimism" about the benefits of screening mammography. Hardly anyone seemed to notice the authors' observation that "the entire difference in mortality between the study and control groups is concentrated among those women who were aged 50 and 59 years at time of death."

In 1973, on the heels of the New York trial, the American Cancer Society and the National Cancer Institute launched an even larger screening program. Rather than study whether widespread mammography screening would reduce the number of annual deaths from breast cancer, the program sought to demonstrate that widespread screening was possible. That it would benefit women was assumed. (Unlike the New York study, which compared outcomes for screened and unscreened women, the 1973 screening program did not include a control group.) Between 1973 and 1980, some 280,000 women aged thirty-five to seventy-four were screened for breast cancer at twenty-nine locations across the United States. While the screening program was under way, and amid concerns that the radiation delivered through mammography might actually increase the rate of breast cancer, the National Institutes of Health convened a panel of experts to take a second look at the New York trial data. The panel concluded that mammography screening could reduce deaths in women older than fifty but said it "found no convincing justification for routine mammographic screening for women under 50 years of age." Again, the caveat was ignored.

Meanwhile, scientists outside the United States were launching their own trials. The most significant were in Canada and Sweden. A Canadian study involving some 90,000 women began in 1980. In Sweden, researchers launched four large trials between 1976 and 1982 involving 283,000 women in five different locations. In all of these trials, the data proved that screening mammography detected more cancers than would have been detected without screening. But the increase in detection didn't always translate into saved lives. As in the New York trial, not all women benefited equally. The Canadian study found that there was no benefit from screening for women aged forty to forty-nine, and even for women older than fifty, the benefit was unclear. The Canadian study has been reviewed repeatedly and interpreted in various ways, but the conclusion after twenty-five years of follow-up was that screening had no impact on mortality for women aged forty to fifty-nine. Two of the Swedish studies that stratified data by age indicated that after eleven and nine years of follow-up, only women aged fifty and older in one trial and aged fifty-five and older in the other saw their risk of dying from breast cancer go down if they were screened. The results of the other Swedish trials found that screening saved lives, but the effect was less pronounced for women under age fifty. The benefits of screening were real, but they had to be understood in a context that was more complicated than researchers had imagined.

Back in America, after two decades of widespread mammography, it was too late for nuance. Unlike some European countries' more centralized health systems, America's breast cancer–screening program was not a program at all. Getting American women to visit the doctor for breast x-rays relied on public campaigns that were intentionally simplistic and easy to understand. "Early detection saves lives" was basic, clear and convincing. Although there was little evidence that screening had a significant impact on breast cancer deaths in women younger than fifty, breast cancer was common enough among fortysomething women that mammography was offered to them anyway. Getting annual mammograms beginning at age forty became a rite of passage for U.S. women. The conversation over when women should begin annual breast cancer screening persisted, but mostly among medical-statistics experts like Diana Petitti.

Many doctors who treated breast cancer patients supported the use of annual mammography beginning at forty. However, a vocal group of epidemiologists maintained that annual breast cancer screening helped fewer women than most people believed and obscured the downsides of mammography, which included radiation exposure and unnecessary testing and treatment. Women, meanwhile, didn't know what to believe, so in 1997 the National Institutes of Health assembled another panel to review all the existing evidence on mammography and hear testimony from experts over a three-day conference in Washington, DC. The twelve-member panel was largely composed of statisticians and medical professionals who did not treat breast cancer patients. The chair was an epidemiologist from Johns Hopkins. Breast cancer was the second leading cause of death for American women in their forties. Perhaps the panel could settle once and for all whether screening mammography should be performed on an annual basis for women in this age group.

In a blockbuster finding, ten members of the panel said the benefits of screening mammography for women aged forty to forty-nine were not supported by evidence. Women who had been dutifully getting their mammograms every year were furious. Radiologists who made their living reading mammograms were in an uproar. The American Cancer Society, which had previously recommended women forty to forty-nine get mammograms every one to two years, revised its guidelines—but in the opposite direction, telling women annual screening was best. Even politicians piled on. Less than two weeks after the NIH panel report, the Senate voted ninety-eight to zero on a resolution urging the National Cancer Institute to ignore the panel's recommendation. President Bill Clinton, whose mother had died of breast cancer, said he supported annual mammography for women in their forties and called on insurance companies to cover the test. Rather than settling the matter, the NIH panel had merely proved that mammography was one of the most contentious topics in modern medicine.

So when Diana Petitti and her colleagues on the U.S. Preventive Services Task Force started looking at mammogram data again in 2008, they knew they were wading into choppy, if familiar, waters. Mammography had become something of a religion, with pink-ribbon reminders seemingly everywhere and mobile mammography vans parked in grocery-store parking lots across the country. To question the logic of fortysomething American women getting annual mammograms was to question faith itself.

More than ever, the pump was primed for outrage. In the summer of 2009, as the task force conducted its analysis, the country was consumed by a fight over whether to enact a sweeping federal overhaul of the American health-care system. (The law had a provision stating that plans funded or subsidized by the government had to cover services recommended by the task force.) The tenor of the debate in America's living rooms and in Washington, DC, reached a nadir in August when members of Congress went home for a recess and were met by angry constituents and protesters. Voters packed town-hall events and shouted down lawmakers accustomed to sparsely attended meet-and-greets. Democrats on Capitol Hill promised that health-care reform would protect consumers and slow the growth in health-care spending that threatened to gobble up nearly one-fifth of the country's gross domestic product. Republican critics, meanwhile, said the reform plan was misguided at best and anti-American at worst, akin to a "government takeover of health care." Some warned the plan would lead to "rationing" of treatment and "death panels" that would allow government bean counters to decide which lives were worth saving.

In November 2009, in the midst of the tumult, the task

force announced its findings. After reviewing decades of data collected on hundreds of thousands of women, the panel concluded that annual mammograms for women aged forty to forty-nine should no longer be routine. Rather, women in this age group should decide with their doctors whether to get screening mammograms. The task force said women fifty to seventy-four should be screened every two years. Mammography could be a lifesaving tool, but saving lives came at a cost, Petitti and her colleagues noted. Screening all women in their forties meant that some lives would be saved, but many more women would be subjected to unnecessary callbacks, biopsies and, in some cases, treatment they did not need. Even if those costs were acceptable, there was another problem. Mammography, the task force noted, wasn't a particularly sensitive test for women forty to forty-nine, whose breasts tended to be denser than those of older women. Mammograms often missed cancers in this group, giving women a false sense of security. On balance, the task force said, most women in their forties without specific risk factors would be better off skipping a screening test that had been promoted for decades as the best protection against death from breast cancer. "I am the ultimate rationalist," Petitti told me. "This is the rational way to interpret this data."

Every major news organization in America covered the release of the new breast cancer–screening guidelines. There were segments on the evening news and cable news and C-SPAN; editorials appeared in newspapers across the country, and there was a flurry of press releases from special-interest groups, including the American Cancer Society and the American College of Radiology. Both organizations said annual mammograms beginning at age forty should remain the norm. Even

Kathleen Sebelius, the secretary of the U.S. Department of Health and Human Services, concurred. Keep doing what you're doing, she told women. It was clear that agreeing with the task force was very bad politics.

As the vice chair of the task force, Petitti had the unfortunate job of being the panel's spokesperson for the new breast cancer guidelines. When she appeared before a congressional committee on December 2, 2009, lawmakers spent more than ninety minutes pontificating on their own experiences and opinions about the benefits and harms of mammography before turning the floor over to the chair of the task force and Petitti. Petitti read from a prepared statement, methodically explaining the task force's thinking, until the chair of the committee cut her off. She had gone over her allotted five minutes. A few weeks later, the U.S. Senate passed a law barring Medicare and Medicaid from refusing to cover screening mammograms for fortysomething women. Democrats promised the task force's guidelines weren't a harbinger of Obamacare rationing, and Republicans promised they were, but politicians of all stripes seemed to agree on one thing—the task force was wrong.

Democratic congresswoman Debbie Wasserman Schultz, who had survived breast cancer, was among the most vocal task force critics on Capitol Hill. The day the breast cancer–screening guidelines were made public, Wasserman Schultz appeared on multiple cable-news shows calling the mammogram recommendations "disturbing" and "inappropriate." To Wasserman Schultz, ending routine annual screening for women in their forties meant that women would be denied the chance to catch breast cancer at its earliest and most treatable stage. "Instead of making things more clear for women," Wasserman Schultz told CNN's Wolf

Blitzer, "these task-force recommendations are making things clear as mud." Appearing on a C-SPAN program the following week, Wasserman Schultz fielded call after call from current and former breast cancer patients. One woman said she was "furious." Another said the idea of eliminating routine screening mammograms for women forty to forty-nine "makes me want to cry." One said her son would be an orphan if not for a routine screening mammogram that found her breast cancer. The government was denying women the health care they deserved. It was sexist, but, more important, it was dangerous.

Wasserman Schultz, diagnosed with breast cancer at forty-one, nodded along, agreeing with each caller. Yes, yes, yes, she said. The irony was that Wasserman Schultz's own breast cancer story was a case study in the shortcomings of mammography for fortysomething women. Mammography, in fact, had failed her.

There is a seductive logic to screening women for breast cancer: Catching a lethal cancer early is better than finding it late. Radiologists can see cancerous tumors on mammograms. Look for breast cancer and you will find it. It seems like a no-brainer. But there are a few inconvenient facts that make this seemingly elegant logic messy.

Some forty million mammograms are performed every year in the United States, but most American women diagnosed with breast cancer discover the disease themselves when they feel a lump or when a gynecologist detects one during an annual exam. Wasserman Schultz found a lump in her breast while showering. She had had a screening mammogram two months earlier that showed no cancer, even though her disease was likely present. When I reached Wasserman Schultz in 2018 to ask about the contradiction between her own experience and her ardent support for annual screening mammograms beginning at forty, she said her mammogram report noted she had a "risk for calcifications," a vague warning that caused her to be more vigilant about examining her own breasts at home. "If you end regular screening mammography for women forty to fifty, more women will die. There is just no question about it," she told me.

It's hard to argue that irradiating women is the surest way to let them know that breast health is important, but when it comes to saving lives, Wasserman Schultz is right. The 2009 task force said that for every 1,904 fortysomething women who get annual mammograms over ten years, one life is saved. A 2018 paper published in the *Journal of the American Medical Association* looked at forty-year-old women specifically and found that if ten thousand of these women are screened annually for ten years, three lives will be saved. There is no doubt that mammography saves lives. The question is whether those lives are worth the trade-offs.

Hard as it is to believe, some breast cancers are not life-threatening, even if they are left alone. Some breast cancers grow so slowly or develop so late in life that women who have them will die of other causes first. Treating these cancers does not extend the lives of women. Treatment simply decreases their quality of life. The more we screen for breast cancer, the greater the number of women who will be treated needlessly, a phenomenon known as *overdiagnosis*. It's hard to avoid this. In many cases, it's still impossible to distinguish between breast cancers that are dangerous and those that are not, so doctors treat them all. Nearly ten times as many women screened annually for ten years starting at age forty will be treated needlessly than will have their lives saved by mammography, according to the *JAMA* paper. Researchers estimate 15 to 30 percent of diagnosed breast cancers across all age groups are treated unnecessarily. False-positive mammograms and unnecessary breast cancer biopsies and treatments cost some four billion dollars a year.

Another complicating truth is that timing matters. Some fast-growing cancers can pop up between annual screenings, making the tool useless. And many slow-growing cancers can be safely detected and treated through a mammography program that screens later and less often, as the U.S. Preventive Services Task Force recommends.

Since the mid-1970s, widespread screening mammography in the United States has increased the rate of women diagnosed with breast cancer by about 30 percent. But the rate of women diagnosed with breast cancer only after it has spread and cannot be cured has barely budged. In our zeal to find as much breast cancer as possible as early as possible, we have overlooked the fact that mammography does not catch all breast cancer and, in fact, some of the cancers it detects are harmless.

About 20 percent of women diagnosed with breast cancer in the United States have what's known as carcinoma in situ. *In situ* is Latin and means "in its original place." Cancers of this type are not life-threatening because they have not escaped into the surrounding breast tissue, known as stroma, on their way to other parts of the body. Yet most women diagnosed with in situ carcinoma of the breast do have surgery. The logic behind removing in situ cancers is that there is a risk they could become invasive. About 15 percent of women diagnosed with in situ cancer have lobular carcinoma in situ (LCIS), abnormal cells inside the breast glands that produce milk. Compared to the general population, a woman with LCIS is at a greater risk of developing invasive breast cancer, although LCIS itself rarely turns into invasive cancer. But most women diagnosed with in situ breast cancer have another type, known as ductal carcinoma in situ (DCIS), which consists of abnormal cells present in the milk ducts. DCIS has been the subject of more controversy than just about any other cancer diagnosis in the modern era.

Doctors have long known about DCIS, often referred to as stage 0 breast cancer. I found a reference to the condition in a 1975 book by breast cancer patient and activist Rose Kushner. A paper published in a major medical journal in 1932 uses the term *carcinoma in situ*. But before the advent of widespread mammography, DCIS accounted for less than 1 percent of all diagnosed breast cancers.

DCIS can't usually be felt. It's most often found when bright white spots show up in clusters on routine mammograms. The white spots are calcium deposits that can be associated with DCIS, invasive cancer or both. Microcalcifications on a mammogram image are often followed by a biopsy in which a hollow needle is inserted into the area to extract tissue that can confirm whether DCIS or invasive cancer is present. Nearly all of those ultimately diagnosed with DCIS, some sixty thousand U.S. women every year, undergo surgery, with most opting for a lumpectomy combined with radiation. About 20 percent of women diagnosed with DCIS have single mastectomies. Some 10 percent have double mastectomies. All of this surgery happens despite the fact that many cases of DCIS will never progress to invasive breast cancer that can threaten a woman's life.

(Some researchers have suggested the share of DCIS that is destined to become invasive cancer is as low as 20 percent.) Doctors do not know precisely which cases of DCIS will progress to invasive cancer because nearly all women with DCIS opt to have it removed immediately, eliminating the ability of researchers to study what would happen if it were left alone. Despite the dramatic increase in DCIS diagnoses and treatment in recent decades, the rate of women diagnosed with invasive breast cancer has not dropped, indicating that the two conditions may not be as closely linked as scientists once believed.

The dearth of good numbers and the knowledge that thousands—and possibly tens of thousands—of women have unnecessary DCIS breast surgery every year has unnerved many in the oncology community. Some wonder if *pre-cancer* is a more appropriate label than *stage 0* cancer. Some have even suggested that the word *cancer* itself might be a misnomer. Hearing the words *breast cancer* can spark in even the most levelheaded woman an urgent desire to remove the threat and ensure it never returns.

Sandra Lee, a television chef and girlfriend of New York State governor Andrew Cuomo, was diagnosed with DCIS in 2015. She initially had a lumpectomy but later opted for a double mastectomy after doctors told her she did not have *clean margins,* a term meaning the outer rim of tissue removed in a lumpectomy contains only normal cells. Appearing on *Good Morning America* to discuss her experience with host Robin Roberts, herself a former breast cancer patient, Lee described her ordeal in dramatic terms. She told Roberts that doctors had said her DCIS was "a ticking time bomb." Lee had been diagnosed in her forties after a routine screening mammogram and said if she had waited until age fifty, as the U.S. Preventive Services Task Force and Diana Petitti recommended, "I probably wouldn't even be sitting here." Lee advised women in their twenties and thirties to be screened for breast cancer and said breast cancer among women in these age groups is "an epidemic." (The probability that a twenty-year-old American woman will develop breast cancer in the next ten years is less than one-tenth of 1 percent. For a thirty-year-old woman, the risk is about half of 1 percent.) "I'm really all over my siblings and my nieces to make sure they get tested. I don't care that my niece is only twenty-three," Lee told Roberts. It's this kind of hyperbole that worries many oncologists. And there are signs that it's having an impact. In a 2017 survey by the Kaiser Family Foundation, 47 percent of nearly three thousand women polled said they believed women should begin screening mammography before the age of forty. As with mammography-screening public messages, it's difficult to find space for nuance when it comes to DCIS. "Women are just so primed that as soon as they hear anything that sounds like breast cancer, they mount this huge response against it," Shelley Hwang, the chief of breast surgery at Duke Cancer Institute, told me.

Pathologists and oncologists now know that not all DCIS is the same. Some is high-grade, with a particular cellular makeup that indicates it has a greater risk of progressing to invasive breast cancer. Some DCIS is fueled by estrogen. Some is not. DCIS discovered in a woman under forty is more dangerous and likely to recur than DCIS found in a sixty-year-old through routine screening. Despite these relatively recent discoveries, treatment for DCIS remains the same for everyone diagnosed with the condition.

Hwang trained to be a doctor in the 1990s. Back then, she said. "It was very black-and-white. You have cancer; you need to take it out. It was very, very easy and just reflected a complete naïveté, a lack of understanding about how complex the disease is." As oncologists have learned more about the wide variations in breast cancer, old paradigms—such as annual mammograms for every woman starting at forty and surgery for all diagnoses—make less sense. "The story of breast cancer, at least in our lifetime, has been very dominated by the messaging around breast cancer, and that's helped to fund a lot of research in this field. It's helped to raise awareness and reduce mortality from breast cancer. But we probably overstepped what is practical, and now we've kind of gone to doing too much," Hwang said. "The unintended consequence is that now women are so anxious about this disease, they're willing to do almost anything within their power to try to avoid ever having it or ever having it again."

In 2016, Hwang and colleagues received a $13.4 million federal grant for a trial they hope will provide some clarity on how to treat DCIS considered to be low-risk. In this study, twelve hundred women are being randomly assigned to either conventional DCIS treatment (lumpectomy with radiation or mastectomy alone) or "active surveillance," basically a watch-and-wait approach. Women in both groups may receive supplemental drug treatment, although not chemotherapy. The results of the study, which will conclude in 2021, could transform treatment for certain types of DCIS, helping women avoid surgery and saving millions of dollars in health-care spending every year.

Doctors are finding more breast cancer than ever, but if the point of screening is to save lives, this is the wrong metric to use. Various studies published on the impact of mammography have found that of the approximately 35 percent decrease in the rate of U.S. breast cancer deaths since 1990, one-third or less of this drop is due to screening. The rest is a result of vast improvements in treatment. When Philip Strax and his colleagues began screening New York City women for breast cancer in 1963, chemotherapy was in its infancy. Other cancer drugs that have dramatically improved the prognosis for women with some of the most aggressive cases of breast cancer were decades away. Genetic testing did not exist. The thinking on how to perform effective breast cancer surgery was in the midst of dramatic change. Consider this: the breast cancer mortality rate for women younger than forty, who are not screened, has fallen nearly 50 percent since 1990.

Four years after Diana Petitti and her colleagues kicked off a new phase in the debate over mammography in the United States, the Swiss Medical Board convened its own panel to review evidence on the topic. The experts concluded that screening mammography's harms outweighed its benefits and said existing breast cancer–screening programs should be allowed to end and no new such programs should be introduced. Two members of the panel published an article in the *New England Journal of Medicine* pointing out an important caveat to studies that suggest that mammography reduces the risk of dying from breast cancer. "We noticed," the panel members wrote, "that the ongoing debate was based on a series of reanalyses of the same, predominantly outdated trials . . . None of these trials were initiated in the era of modern breast-cancer treatment, which has dramatically improved the prognosis of women with breast cancer." The panel members also cited research indicating that women dramatically overestimate not only the benefits of screening but also their risk of dying of breast cancer. As treatment continues to improve, the usefulness of widespread population screening shrinks even further. A 2013 meta-analysis of mammogram trials involving some six hundred thousand women found that breast cancer screening reduces breast cancer mortality by 19 percent; among women aged thirty-nine to forty-nine, the reduction was just 8 percent. But when the authors of the analysis looked only at trials with adequate randomization, they found that screening did not have a statistically significant effect on breast cancer mortality.

Part of the problem is that mammography is a somewhat crude method of detecting breast cancer. X-rays provide an image of the anatomy of a breast but little else. In the modern era, the biology of a breast cancer tumor is often more consequential than how large it is when found. Some types of breast cancers are less aggressive and can be easily and effectively treated even if the tumors are very large and have been present for a long time. Other harder-to-kill types of breast tumors can be only a few millimeters across and still spread around the body quickly.

The technology of mammography is limiting in another way. A mammographic image is gray with streaks and clusters of white. It looks remarkably like the surface of the moon, and dense breast tissue, common in younger women, can make tumors impossible to see. (A mammogram performed on my own dense breasts after I developed symptoms suspicious for breast cancer showed no invasive cancer, even though the disease was present.) Some researchers have said using mammography to find cancer in a dense breast is like looking for a golf ball in a snowstorm. This is partly why the authors of the 2018 *JAMA* paper estimated that within a group of ten thousand forty-year-old women screened annually for ten years, thirty-two will die of breast cancer regardless of the screening—more than ten times the number of women whose lives will be saved.

The truth is that there is nothing magical or precise about the age of forty, or fifty, or fifty-five. What happens if we go a step further? Morning-show chatter notwithstanding, even the most ardent mammogram evangelists have stopped short of suggesting that widespread screening should begin at age thirty. Why not start then? Radiologists would find more breast cancers earlier and save more lives, but the collateral damage in the form of unneeded radiation, biopsies and treatment would be huge. While rooted in statistics, the decision on when to begin screening is subjective, which is what the U.S. Preventive Services Task Force members, including Diana Petitti, said in 2009—not that all women in their forties should skip mammograms, but that they should decide for themselves whether they are willing to risk a statistically greater chance of being harmed unnecessarily in order to gain a statistically smaller benefit of being helped.

The 2009 task force guidelines noted that screening all women forty and older every year has emotional consequences as well as physical ones. Suspicious mammograms and biopsies that turn out not to be cancer cause unnecessary worry and anxiety. "The arguments made are that there are false positives and it's scary. They literally use those terms," Wasserman Schultz told me, adding that she found it sexist for a government task force to cite women's emotional well-being as a reason not to screen them for breast cancer. "I'm just not willing to write off the lives of women forty to fifty because we are worried about affecting their psyche."

Rather than sexist, the task force's message could be empowering. For centuries, female patients lacked agency over their health, which was managed by seemingly omnipotent male physicians. On the whole, women were not encouraged to take control over their bodies or to ask questions of the doctors who cared for them. With its 2009 breast cancer screening guidelines, the U.S. Preventive Services Task Force was advising women to get in the driver's seat and educate themselves about the nuances of mammography so they could make their own choices about whether and how to use the tool. Petitti's numbers were real, but they never stood a chance. Breast cancer can cause psychological trauma for patients, their families and even medical experts. Heart disease kills seven times as many American women as breast cancer does, but it does not threaten womanhood, motherhood and femininity. It does not usually strike with indiscriminate ferocity and leave women bald and in bodies that will never look or feel the same. Breast cancer is emotional and it makes women and doctors emotional. In the face of so much feeling, statistics on costs and benefits can seem like an annoyance rather than a reality check.

In 2016, Diana Petitti was no longer on the task force, but its new members repeated the same breast cancer screening recommendation as in 2009, advising average-risk women to get biannual mammograms starting at age fifty and fortysomething women to weigh the pros and cons in their decision about when to begin screening. This is confusing and unrealistic, according to some physicians I talked to. (A 2017 study published by the *Journal of the American Medical Association* found that 80 percent of primary-care doctors and gynecologists still recommend women begin breast cancer screening at forty.) A woman who is told she doesn't have breast cancer after getting an unnecessary biopsy is not angry; she is relieved, these physicians say. But there may be unintended consequences. A meta-analysis published in the journal *Health Technology Assessment* in 2013 reported that when mammogram findings that are suspicious for breast cancer are later determined to be benign, the associated psychological stress of waiting for a diagnosis can last for up to three years and reduce the willingness of women to submit to future screening. A 2014 study of about one thousand women published in *JAMA Internal Medicine* noted just the opposite; the authors wrote that stress from false-positive mammograms was short term and that these women were more likely to undergo future screening than women who were given the all-clear after their mammograms. In any case, some primary-care doctors and gynecologists charged with talking to women about breast cancer screening do not understand the complicated data analyses that exist for mammography, let alone have the time to convey such nuance to their patients.

In this vacuum, women rely on one of the most effective public-health message campaigns in history. In general, the decades-old mantra "Early detection saves lives" has given American women, through no fault of their own, a warped sense of the efficacy of mammography.

Reading the historical medical literature on breast cancer screening and interviewing experts like Diana Petitti made me wonder if there wasn't a way to both acknowledge the fear women feel about breast cancer and educate them enough that they could make fully informed decisions on screening mammography. For example, what if a physician said this to a woman who turns forty and qualifies for an insurance-covered annual screening mammogram?

> *There is a 1.5 percent chance you will be diagnosed with breast cancer in the next ten years. If a cancer is detected, there is a 20 percent chance it will never hurt you, but you will be treated anyway. Do you want a mammogram?*

Lost in the discussion about screening mammography and overdiagnosis is an even more pressing problem. If thirty-two forty-year-old women out of ten thousand screened die of breast cancer anyway, mammography is failing in another way. What about cancers that mammography does not find? What about *under*diagnosis? If the answer is not to search as hard for breast cancer, then what about the forty thousand American women who die every year of the disease? Some of these women did not respond to existing treatments, but most were diagnosed too late, even if they didn't know it at the time. Even as physicians diagnose too many of the wrong cancers, they are failing to diagnose enough of the right cancers, the kind that will kill and need to be caught early. What if instead of not searching as hard for cancer, we just searched smarter?

2

CSI: breasts

The thing to know about breast cancer doctor Laura Esserman is that she doesn't see problems. She sees systems. To Esserman, almost anything that's going wrong is part of a larger equation that has interlocking parts and competing priorities. You can't solve a problem unless you change the system. The other thing to know about Esserman is that often she believes she is the ideal person to enact the kind of systemic change that's needed. "People think I'm too outspoken and too disruptive. Disruptive people are annoying—they make you change things and do things and that's hard," Esserman told me when we met for coffee in the fall of 2017 near her home in San Francisco's Haight-Ashbury neighborhood.

In the community of breast cancer professionals in America, Esserman is both famous and infamous. Her assertiveness and desire to upend existing systems of breast cancer diagnosis and treatment elicit extreme reactions in those who know her work best. One top oncologist told me Esserman was "dangerous." Another said she "has a big name" and therefore gobbles up a lot of federal grant money that could be better spent elsewhere. Others were highly complimentary and said Esserman was provocative in the best way—forcing fellow cancer doctors to reevaluate their methods and beliefs in service of delivering better care.

Esserman's big name has come from her voluminous publication record and membership on important national breast cancer committees. But Esserman's reputation has also grown because she speaks her mind and is a media darling. When the *New York Times* came calling in 2015, eventually publishing a flattering profile headlined "A Breast Cancer Surgeon Who Keeps Challenging the Status Quo," Esserman deboned a salmon with surgical tools, mashed potatoes and sautéed arugula for the writer of the piece who had come over for a dinner party. Esserman's "culinary magic," the article gushed, "occurred while she steered her colleagues through a gamut of complex topics, including the longstanding debate over breast cancer screening." The cynic in me wondered if Esserman had put on a show for the *Times* writer, knowing that such a display was bound to end up in print, portraying her as a polymath who could cure cancer *and* cook a delicious meal at the same time.

But when I met Esserman in the fall of 2017, with no salmon in sight, I felt sucked into her vortex too. Wearing an off-the-shoulder blouse and large black and brass hoop earrings, Esserman shifted between leaning almost uncomfortably close to me and arching back with her elbow perched on the back of her chair. She told me she thought that half of all women probably did not need to be screened for breast cancer. She talked about competitive, critical doctors who she knew would be "in the way" when she proposed changes. She talked about how a lot of problems in cancer research are based on "the way we assign credit." Of doctors who bristled at Esserman's call for less radiation treatment: "Oh my God. You'd think I was taking away their firstborn!" She told me she was "on a mission" to reform all of health care, not just breast cancer.

Esserman is the youngest of four children and grew up in Florida. Her father was a car dealer. Her mother dreamed of becoming a professional writer but struggled to find that kind of work as a married woman in the 1950s and 1960s. She became a teacher instead and made sure her children understood that the obstacles she faced as a working woman weren't just her own problems. They were baked into postwar America, part of a political, economic and social system designed to keep women from achieving the same success as men.

The message stuck. Esserman was valedictorian of her 1974 high-school class and gave a graduation speech arguing for the adoption of the Equal Rights Amendment, an ultimately unsuccessful political effort to make gender discrimination illegal under the U.S. Constitution. Esserman's high-school principal called her mother the next day complaining about the controversial subject matter of the speech. "Don't you think she's earned the right to say whatever she thought was important?" Esserman's mother replied.

Esserman graduated from Harvard and was accepted to the Stanford University School of Medicine. She paid part of her tuition by working as a teaching assistant for David M. Eddy, a physician and an applied mathematics professor. Eddy was conducting research on Pap smears, developing mathematical models to determine how frequently women needed to be screened for cervical cancer. Standard medical practice back then was to screen women every year, but Eddy's work demonstrated that screening every two or three years was equally effective. Eddy's evidence was clear, but practicing physicians were slow to adapt. Esserman remembers one Stanford obstetrician-gynecologist was particularly resistant, saying his patients couldn't keep track of a years-long screening schedule. "I was just like, 'Oh my God, nobody gives a shit!'" Esserman said. She was incensed, particularly because there was a shortage of qualified experts who could accurately analyze Pap smear results, a problem exacerbated by the annual exams. "We were doing more harm, and nobody cared," said Esserman. She started searching the halls of Stanford for more doctors like Eddy, people who were unafraid to challenge conventional wisdom.

Along the way, Esserman became a chief resident in the surgery department. One of her responsibilities was to book guest speakers to address medical students and other residents on the latest developments in medical practice. One of the speakers Esserman invited was a Stanford professor named Alain Enthoven. Enthoven was an economist in the business school, not a doctor, but he had written a book about the high cost of medical care, and Esserman thought the future doctors in her charge should hear his perspective. After Enthoven's presentation, he asked Esserman to drop by his office.

Enthoven had noticed that Esserman thought differently than most doctors. Just as her mother had realized that the disadvantages she faced as a woman were part of a system of discrimination, Esserman rightly saw that many of the flaws in health care were far more complicated than they seemed. Enthoven had a proposal: Put your medical career on the back burner for a few years, he told Esserman, and get a business degree. It was all worked out, he said. A fellowship had been secured to pay her tuition—all she had to do was fill out some paperwork. At first, the idea seemed like madness. In addition to being a chief resident, Esserman was pregnant with her first child and immersed in research on cancer-fighting antibodies. "I said, 'Alain, I've been training for ten years now. I just did this immunology fellowship. I can't imagine doing this,'" Esserman said. "I was flattered, of course, but I couldn't." But Enthoven didn't give up. Every so often, he'd call Esserman. "I don't have your application yet," he'd say.

Finally, she relented. Maybe business school did make sense. In her decade of medical training, Esserman had seen the American health-care system up close. The more she learned, the more frustrated she became. There were the hospital administrators who brushed aside better policies because they were difficult to manage. There were doctors more dedicated to their disciplines and careers than to their patients, staking out turf and defending their domains rather than responding to new science and working together. Maybe learning about business would give her the skills and credibility she needed to outflank her colleagues and bosses.

So Esserman took Enthoven up on his offer, juggling a part-time clinical practice and new motherhood with business courses. Esserman quickly realized that the principles she was learning could be applied to medicine in countless ways. Organizational behavior courses instilled in Esserman the value of teamwork. Marketing courses helped her sell her ideas to patients and colleagues. Accounting classes taught Esserman how to read the financial documents she might one day review as a hospital administrator. And microeconomics taught her the principle of diminishing marginal returns, the idea that doing more of something can lead to less and less benefit over time.

This last lesson would prove particularly useful when Esserman was hired to be director of the Carol Franc Buck Breast Care Center at the University of California, San Francisco, a position she has held since 1997. From that perch, she has thrown bombs at the conventional wisdom around breast cancer screening and standard treatments. Her voice, perhaps more than any other, has helped shift the modern conversation about the disease. Esserman has said for years that DCIS should not be called cancer. She has been the lead author on published papers stating that more detection of early breast cancer through screening "is not necessarily beneficial." She has written that "early detection may not be the solution for aggressive cancers" and that to reduce breast cancer mortality, "the first step is a change in mindset in scientific discovery efforts and clinical practice." Esserman said the dramatic uptick in breast cancer diagnoses due to mammogram screening "adds the burden of diagnosis to hundreds of thousands and engenders needless fear" and that "we need to curb the urge to intervene with more thought about what is truly valuable." Perhaps most controversial, Esserman and coauthors pointed out that the U.S. Preventive Services Task Force breast cancer screening recommendations could cut overall screening costs by more than half. (Approximately eight billion dollars a year is currently spent on breast cancer screening in the United States.)

In the past two decades, knowledge about breast cancer risk and improvements in treatment have transformed what it means to be a breast cancer patient in America. And yet, the basic

U.S. screening paradigm—a mammogram every year or two for all women over forty or fifty—has remained the same, in part because the data informing opinions on screening was generated by large-scale studies conducted before recent advances. If the firestorm over the 2009 U.S. Preventive Services Task Force guidelines demonstrated anything, it was that parsing the same old data sets in different ways is unlikely to shift public or scientific opinion on screening in any meaningful way. We need new data. In addition to writing about the need for reform in breast cancer screening, Esserman is running one of the largest and most important screening trials in the modern era, hoping to inject some fresh evidence into the debate.

Called WISDOM (Women Informed to Screen Depending on Measures of Risk), the trial will do something that's never been done on the scale needed to finally modernize breast cancer–screening norms in the United States. It will study whether women are better off being screened every year beginning at forty or following a personalized screening program designed around their individual characteristics. Just as breast cancer treatment is becoming increasingly precise, Esserman thinks screening should be too. WISDOM will prove if she is right.

Ninety percent of women do not know their actual risk of developing breast cancer, according to research cited by Esserman. Women in the risk-guided arm of WISDOM will provide their family and personal medical histories and be screened for genetic mutations known to increase a woman's odds of developing breast cancer. Women found to be at the highest risk for breast cancer will be screened with magnetic resonance imaging (MRI) or mammography every six months. They may also be prescribed hormone-related drugs or undergo prophylactic surgery to reduce their risk. Women who have the lowest risk of developing breast cancer will get a mammogram every two years.

Using all that physicians have learned about breast cancer risk to tailor a woman's individual screening schedule is so logical, one has to wonder why the idea has not been tried before. In small ways, it has. Women who test positive for *BRCA*, an inherited genetic defect that can dramatically increase the chances of developing breast cancer, are typically advised to undergo earlier and more frequent screening or opt for prophylactic surgery. What makes WISDOM unique is that it goes beyond *BRCA;* it takes into account other genes, plus a woman's family history and personal health status, including her age at the onset of menstruation and menopause, pregnancy history, breast density, race and whether she has ever had radiation treatment known to increase the risk of breast cancer. It's a systemic approach rather than the one-size-fits-all paradigm that has guided screening for more than fifty years. Esserman secured a $14.1 million federal grant, funded through the Affordable Care Act, to launch the trial. She's hoping that the study results will generate a large return on this investment, saving insurers and patients money by screening some women less often and by identifying high-risk women before they develop late-stage breast cancer.

When I spoke to Esserman again in October 2018, she admitted the trial had gotten off to a slow start. Not everyone, it seemed, was as excited as she was to create a new system for mammogram screening. Insurance companies, which Esserman hoped would contribute funding to the study, had mostly balked. Blue Shield of California signed on to the project early and remained on board. "I had thought the insurance industry would see why this makes so much sense," Esserman told me. "But I don't think they see it as their job to have a role in making sure we have better data, better value." Accrual to the trial had also been slow, with only 16,500 out of the 100,000 needed enrolled two years into the five-year study. And when we talked, just two health systems had agreed to participate, the University of California and Sanford Health, located in the Midwest. Esserman said more hospitals would be signing on.

In the end, the trial may even identify women who do not need to be screened for breast cancer at all. These are existing, but as yet unidentified, women who are not helped by screening, but are harmed by unneeded biopsies and treatments. "I think that's quite likely and quite possible. That should be an important goal," said Esserman. She is hoping the size and design of WISDOM will produce results that even the most committed annual-screening evangelists can accept.

The upper echelon of the American oncology community is full of big egos, but most doctors are politic when discussing medical tests, treatments or colleagues they disagree with. For a long time, Esserman was not. She told me she had a more abrasive approach early in her career and occasionally was kicked off or not invited back to various cancer committees. After Debbie Wasserman Schultz started her 2009 media tour criticizing the task force breast cancer–screening guidelines, Esserman called the congresswoman to confront her. "Basically, she hung up on me," Esserman said.

In recent years, though Esserman has kept the same views on screening and treatment for breast cancer, she has changed the way she talks about them. "Here's the thing," she said, leaning in. "People who have been treated in the past don't want to hear that they got the wrong thing. How does that help them? It's just hurtful and mean." Instead, Esserman says breast cancer doctors need to admit they made mistakes and do something that's rare in the oncology community—apologize. "There were a lot of things I didn't know, and my outcomes weren't as good as they are today," said Esserman. "I'm sorry. I wish I could go back and do it all over, but I can't. But I've learned from it."

In the coming decade, it is possible that breast cancer screening as we know it will come to an end, replaced by a system that takes into account each woman's personal risk of developing the disease. If history is any guide, there will be resistance to this change from women and doctors. But there is more reason than ever to expect a paradigm shift, in part because it's not just a polarizing figure like Laura Esserman calling for change.

On the other side of the country, Etta Pisano, a Harvard professor and chief research officer at the American College of Radiology, is leading a second large-scale screening trial based on the idea of personalization. Pisano was part of a group that helped develop guidelines for the country's first mammography law in the 1990s regulating the quality of the machines used to x-ray breasts. Later, she chaired a large study that proved that digital mammography worked as well as, and sometimes better than, film mammography, paving the way for adoption of the new technology nationwide.

Like most radiologists, Pisano believes the benefits of screening mammography outweigh the harms. But she also knows that the test is flawed, that it misses some cancers and occasionally indicates the presence of disease in masses that turn out to be harmless. Sometimes, mammography correctly identifies a breast cancer tumor but the patient doesn't follow up on the information, making one wonder if there was any point in doing the test at all.

In 1989, when Pisano was a young doctor, she took a job as a radiologist at the University of North Carolina, Chapel Hill, where she became the chief of breast imaging. In that post, Pisano spearheaded an initiative to increase the screening-mammography rate among women living in rural areas. These groups had screening rates far below their counterparts' in the city of Chapel Hill. To reach the underserved women, Pisano managed a van equipped with a mammogram machine that could be driven into the countryside. The program was a success. With the mammogram van, UNC Chapel Hill technicians screened a large population of women who had rarely or never had mammograms before, finding many breast tumors, some of them quite large. But when Pisano went back to review the overall results of the mammogram-van project, she felt deflated. "There was a pathetically bad follow-up," she said. About two-thirds of the women UNC radiologists suspected had breast cancer never had biopsies or other follow-up care to confirm the diagnoses or begin treatment. "The women were disenfranchised from health care," Pisano said. "They weren't getting workups because they were not connected to the health system." The experience taught Pisano what Esserman had learned in business school—that interconnected problems are hard to solve without new systems.

Pisano is the head of a study called TMIST (Tomosynthesis Mammographic Imaging Screening Trial), which is focused on a relatively new technology called 3-D mammography. Like standard digital mammography, 3-D relies on x-rays, but it adds a computer program that merges two-dimensional digital images into a three-dimensional picture of the breast. More than one-third of all mammography machines in the United States are 3-D units, which cost as much as 50 percent more than their 2-D counterparts. Studies have shown that 3-D mammography finds more breast cancers, particularly in women with dense breasts, but it is unknown whether the technology achieves the ultimate goal of better screening, which is saving more lives. Like WISDOM, the TMIST study will assess participants' risk of breast cancer—examining age, breast density, family medical history and genes—and assign each woman in the study to undergo screening every year or every other year based on such risk factors. The goal is to determine whether 3-D mammography can identify more of the cancers that are likely to spread and become lethal without increasing one of the problems inherent in standard mammography—overdiagnosis of non-life-threatening cancers. The study will include 165,000 women and wrap up around 2025.

Most participants in Pisano's study will provide blood and DNA samples in hopes that researchers might one day identify genetic or other markers that indicate a higher breast cancer risk—or even the presence of cancer. Several well-funded companies are racing to develop "liquid-biopsy" technology, in which the presence of cancer can be identified through tests. These companies are hoping such tests might one day be performed alongside screening mammography. One venture, a Silicon Valley start-up called Grail, announced in 2017 that it was enrolling one hundred twenty thousand women in a clinical trial that would match women's screening mammography results with analyses of their blood samples. Preliminary Grail data from a different research project involving blood tests for multiple types of cancer was presented at a major cancer conference in 2018. The data showed that the company's breast cancer screening tests, designed to detect DNA shed from malignant tumors and circulating in the bloodstream, were not ready for prime time. Researchers reported the tests failed to detect many cases of breast cancer in women already diagnosed. Following the results, Grail decided to amend its breast cancer screening trial to test for multiple types of cancer. Liquid-biopsy technology might one day transform cancer screening, but it must overcome some major hurdles first. Critics of screening mammography worry that, rather than solving the problems of overdiagnosis and underdetection, the testing could exacerbate both, returning even more false positives (increasing the number of women who undergo unnecessary testing or treatment) or false negatives (giving women a false sense of security). Another leading edge of breast cancer screening is artificial intelligence. Radiologists already read mammograms with the help of computer software. Scientists are studying whether more sophisticated programs might be able to read mammographic images as well as or more accurately than humans.

Due to the efforts of people like Esserman and Pisano and epidemiologists like Diana Petitti, the debate over screening mammography will soon be, if not over, then more informed. "We're in the dark ages in the way that we screen for breast cancer," Elizabeth Morris, the chief of breast imaging at Memorial Sloan Kettering Cancer Center, told me. "The way that we do things now, which is sort of population-based, generic, everyone gets the same thing, is going to change radically."

The studies run by Esserman and Pisano will inject some much-needed fresh data into the screening mammography conundrum and, hopefully, reduce the amount of unnecessary pain and worry women have experienced since the technology came online as a tool. But neither trial is likely to dramatically increase detection of the worst, life-threatening breast cancers that are obscured on mammograms or form and grow too quickly to be found early through an annual or biannual mammogram screening program. These are the cancers that, most days, occupy the mind of Dr. Christiane Kuhl. A German radiologist and chair of the radiology department at RWTH Aachen University, Kuhl wants to find the worst breast cancers that aren't picked up by mammograms. To do this, Kuhl is studying the feasibility of using a tool that has existed for nearly forty years but has not played a central role in population-based breast cancer screening.

Magnetic resonance imaging (MRI) uses a powerful magnet to create images of the body. Hydrogen nuclei (protons) become aligned in a magnetic field, so first, the MRI magnet aligns the direction of the protons' spins. Next, the machine emits radio waves in pulses, which causes the hydrogen protons to change their spin direction. When the current is turned off, the protons return to their previous states. Since hydrogen is so abundant in the human body, an MRI is a remarkably effective tool for mapping our insides. Unlike a mammogram, which is an x-ray that captures an anatomical image, an MRI highlights abnormalities within tissues. When used with the IV contrast agent gadolinium, an MRI can also create a picture of blood flow. Because tumors need blood to survive, MRI is particularly good at finding cancer in breasts, even breasts that are so dense that tumors can hide in mammographic images.

Breast MRI has been in use since the 1980s and is already utilized, on a small scale, as a screening and diagnostic tool in breast cancer. Women known to be at very high risk of developing breast cancer are sometimes scanned with MRIs, and the technology is frequently used to confirm cases of breast cancer. But Kuhl and others have amassed data indicating that MRI could play a role in population-based screening as well. In a 2007 paper published in the *Lancet,* Kuhl and colleagues found that MRI was better than mammography at picking up cases of "high-grade DCIS," the type most likely to turn into aggressive invasive cancer. Finding these cancers early is critical. Studies also show that MRI is better than mammography at detecting the most dangerous, fast-growing invasive breast cancers. These tumors tend to be highly "angiogenic," meaning they rapidly form new blood vessels, which makes them highly visible on an MRI image. MRI also uses no radiation.

There is no question that MRI is better at finding invasive breast cancer than mammography is, particularly in women with dense breasts. What is not yet known is whether women at average risk of developing breast cancer would be better off being screened with MRI and whether the expense and inconvenience of the test are worth its increased sensitivity. A breast MRI in the United States costs far more than a standard digital screening mammogram. The test also requires a patient to get an IV and lie inside a loud MRI machine, a far more elaborate ordeal than a simple mammogram, which takes much less time and requires no IV. MRI may also be *too* good at finding breast cancer and so could exacerbate the problem of overdiagnosis.

Kuhl is working through these problems, studying whether they could be mitigated enough to justify using MRI as a widespread breast cancer–screening tool. She has homed in on something called "abbreviated MRI," which is much faster and cheaper than its conventional counterpart. A standard thirty-minute MRI creates a long series of images that a radiologist might spend twenty minutes reading. By contrast, an abbreviated MRI scan takes less than five minutes and produces just a few images that can be read by a radiologist in seconds. It is also far less costly than a conventional MRI. In feasibility studies involving hundreds of women, Kuhl and colleagues have found that abbreviated MRI is just as accurate as conventional MRI when used to screen for breast cancer. Elizabeth Morris, the Sloan Kettering radiologist, told me that conventional MRI machines and procedures are overpowered when it comes to breast cancer screening. "The really smart people go into [the field of MRI] and then just to show how smart they are, they do all these special sequences," she said. "It becomes this very boutique-y, long exam with like ten sequences when maybe only once in a while you might need that sequence."

Kuhl's studies have also suggested that screening women every two or three years with abbreviated MRI eliminates the problem of "interval cancers," tumors that develop between screenings or are present but can't be seen on mammogram images. Because MRI is more sensitive and picks up tiny and fast-growing cancers very early, the scan finds some cancers that are not detected by mammography. Kuhl admits that more study is needed, but if her early findings are correct, it could mean that MRI can detect breast cancers far sooner than any technology program in history, achieving the original aim of breast cancer screening—finding dangerous breast cancer early enough that it can be defeated.

The idea of screening millions of women with breast MRI, even an abbreviated version, unnerves many doctors, who say it's far too early to know whether the upside of the test (finding more cancer) justifies the downside (finding too much cancer). Another issue brought up by many doctors I spoke to is that the contrast agent gadolinium, a heavy metal that the Food and Drug Administration says is safe, could have unforeseen effects if used more widely and frequently. Still, most physicians I asked said they are intrigued by the idea of abbreviated MRI. In addition to Kuhl's work, a U.S.-based study and several in Europe are currently under way to further investigate the technology as a screening tool.

"In my personal opinion, people who say we can't use MRI due to lack of evidence are the ones who simply don't want a change," Kuhl told me. "Of course, overdiagnosis is a problem," she said. "But the main problem that we have right now is underdiagnosis. If you set up a screening test, the one thing that I would expect as a woman is, if I do have a cancer, then that cancer will be found. Not diagnosing cancer that has the potential to kill is clearly the more important mistake than finding a cancer that may not kill." Kuhl said she is watching Esserman's WISDOM study, which is testing personalized screening. "I think it's a good approach because compared with the 1970s"—when the first mammogram trials were conducted—"we have so much more knowledge about risk stratification... But still, the answer is the same. Mammography, mammography, mammography."

Kuhl said questioning the value of mammography, as she is doing through her research and in the pages of top medical journals, is seen as "heresy." "The belief in mammography, at least in the U.S., is an almost religious belief," she said. The oncology community's obsession with mammography has saved lives, but as I talked to Kuhl it became clear to me that faith in the technology has also stunted scientific inquiry. Other tests exist or could be developed to improve breast cancer screening, but mammography has remained the dominant technology, even in the face of evidence that it falls short for many women. For all the awareness about breast cancer, women, by and large, remain unaware of their risk of developing the disease and oblivious to the fact that the primary tool used to screen for it often does not work.

Having dense breast tissue can hide a tumor on a mammogram and also increases the odds a woman will develop breast cancer. And yet, for decades, women with dense breasts—nearly half of all adult women and an even larger share of younger women—were not told that the value of their annual screening mammograms was diminished by this fact. Women might still be in the dark if not for a Connecticut woman named Nancy Cappello. Diagnosed in 2004 with advanced breast cancer just six weeks after a supposedly clear mammogram, Cappello learned she had dense breast tissue, which likely obscured her tumor on the screening test. After undergoing treatment, Cappello founded a nonprofit advocacy organization called Are

You Dense? and started lobbying states and the federal government to pass laws requiring that mammogram results include information about breast density. Cappello died in 2018 of complications related to a rare blood cancer she developed due to the radiation and chemotherapy she received after her 2004 diagnosis. Her effort to inform women about breast density has been a huge success. At least thirty-five states now require women to be informed about their breast density after they undergo screening mammograms. (In March 2019, the Food and Drug Administration proposed a rule that would expand such policies nationwide.) The information empowers women to talk to their doctors about whether further screening using other methods might be warranted. Sometimes this means MRI. More frequently, though, women at high risk of developing breast cancer are scanned with ultrasound.

Ultrasound, which uses sound waves to scan tissue, is relatively cheap and fast and can help radiologists determine, for example, whether a suspicious mass is solid or filled with fluid. Fluid-filled masses are usually cysts. Solid masses tend to be cancer. Studies have shown that adding ultrasound to a mammogram screening program increases the number of cancers detected.

"I had no concept of the profound limitations of mammography until it began to manifest in my own clinical practice," said Deborah Rhodes, an internist at the Mayo Clinic in Rochester, Minnesota. Like many primary-care physicians, Rhodes was tasked with sending women for mammography screening and explaining the results. Over the years, she noticed that some of her patients were diagnosed with relatively large breast cancer tumors even though they had undergone annual screening that revealed no tumors. Many of these patients had dense breasts and had been falsely reassured by mammogram results that showed nothing. "If you really said to them, 'You can participate in this test every year, but be aware that if you have dense breast tissue, the mammogram is more likely to miss it than find it'—if you really communicated that in plain English, what decision would that yield?" said Rhodes. "The mammography community and advocacy groups have been very reluctant to communicate it in plain English precisely because they worry that it will erode confidence in mammography and women won't do it at all. So then whatever marginal benefit you gain from it is going to be completely lost. But the solution cannot be to withhold that information. The solution has to be that we need to find a better way to screen. The problem with the history of mammography is that they've made these tiny, tiny, tiny gains and there's never been a quantum leap. We need a quantum leap."

Awareness about breast density has increased calls for new screening tools. To that end, Rhodes is studying yet another technology, called molecular breast imaging (MBI). Studies suggest the test may be better than mammography at detecting small invasive tumors in dense breast tissue. Women are injected with a radioactive tracer that is absorbed by the body, especially tissues that have a lot of rapidly dividing cells, such as tumors. Once the tracer is coursing through a woman's bloodstream, she is scanned using a machine similar to a mammogram apparatus that scans her breasts between two plates. The cancerous cells in the breast absorb more of the radioactive tracer than normal cells do, so on the image, areas of malignancy appear as dark spots. MBI is relatively new and not widely available, but Rhodes hopes it might one day be added to the imaging tests available to screen for breast cancer nationwide. A study at Mayo is enrolling three thousand women and comparing the breast cancer detection rate of MBI and 3-D mammography. Rhodes told me that a preliminary analysis conducted halfway through the study found that MBI detects three times as many invasive breast cancers as 3-D mammography. "I'm not saying mammography is useless and that it should be abandoned," Rhodes said, "but it's not enough."

3

diagnosis

There's an avocado tree in the backyard!"

I didn't need a sales pitch. The idea of moving to California was already lodged in my brain, the backdrop for every daydream I had about the future. In the weeks after my husband, Collin, applied for a job in Los Angeles, I spent most evenings in the tiny garden outside our basement apartment in Brooklyn thinking about palm trees and parking lots with plenty of spaces.

Space was something I had longed for ever since I left my family's farm in upstate New York and moved to the city for graduate school. Our farmhouse was surrounded by alfalfa and cornfields, wide-open sky in every direction. In the summertime, my mother grew irises along the barn, and we had a rhubarb patch in the yard, just past the clothesline. When the days grew short in wintertime, we shoveled snow until the insides of our nostrils froze. After a decade in New York City, I desperately missed privacy and quiet. I felt choked by the small corners of the big city, even as I was grateful for New

York's bounty—a brilliant, worldly circle of friends and a career in national magazine journalism that had once seemed out of reach for a country girl like me.

Truthfully, I didn't have much to complain about. An internet-fueled revolution in the news business had left many of our journalism-school friends laid off or eager for careers in more stable industries. But Collin and I had both managed to hang on. In the fall of 2013, as a staff writer for *Time* magazine, I was working on my fourth cover story and teaching journalism at Columbia University, our alma mater. Collin was creating and managing shows for the city's public radio station, WNYC. We were far from wealthy, but we earned enough to live on a leafy street in Brooklyn and shop at the farmers' market on Saturdays. Best of all, we could afford day care.

Collin and I had conceived Evangeline easily, and the pregnancy raced by. Evie started sleeping through the night when she was a few months old and grew into a thriving, precocious toddler. She wore a set of green fairy wings around our apartment and loved peas and corn chowder.

When Collin flew to Los Angeles to interview for a management job at a public radio station, Evie and I tagged along. I wanted to gut-check the certainty I felt in my bones that a cross-country move was the right one. We took Evie to Santa Monica's Third Street Promenade near the beach and she danced to "Hotel California" played by street musicians, her red-sandaled feet planted firmly on top of Collin's.

A few months later, job offer in hand, Collin loaded our dog, Lola, into the car and drove out to Los Angeles to look for a place for us to live. He settled on a two-bedroom bungalow nestled in the city's hip northeast quadrant. It had a fireplace, deck, gleaming new kitchen and a long driveway. In the chilly October weeks before Evie and I left Brooklyn, Collin e-mailed me forecasts for Los Angeles, all sun and seventy-degree days. He sent me top-ten lists of best hikes in Southern California and blog posts about restaurants near our soon-to-be home. A week before we arrived, he e-mailed a picture of the sun setting over the Pacific. *Ready for this?* said the subject line. I e-mailed him back a picture of Evie's stroller shoved inside a crowded subway car filled with exhausted commuters. My subject line: *Never again.*

The day Evie and I arrived, Collin picked us up at the airport and drove us to Malibu so we could see the blue Pacific in the early-evening light, sparkling and stretching westward. We kept the windows down and drove through the canyons to our new home. I walked in the front door and went past the sea of boxes I had carefully packed back in Brooklyn. There it was. An avocado tree, its leaves a deep green.

That night, Collin and I sat on the kitchen floor, drank California craft beers and toasted our new life. A few days later, I had gotten Evie down for an afternoon nap and was putting dishes away in the kitchen when a chat window popped up on my open laptop. It was Collin, writing from work.

Nothing but up from here!

A cancer diagnosis does not usually happen all at once. Rather than having a single somber conversation with an oncologist, a patient generally learns the truth through a trickle of test results and expert opinions. Sometimes, it starts with a mystery: A cough that won't go away. An upset stomach. A funny-looking freckle. A lump found in the shower. I never felt a lump, nor did any of the highly qualified breast cancer doctors who examined me. The reason I found myself being squeezed and palpated by these doctors was that my breast cancer had announced its presence all on its own. It had started oozing.

Breasts are never really the same after you've used them to nourish a child. In the summer of 2014, nearly a year after we moved to California and two years after I had stopped nursing Evie, my left nipple began occasionally discharging a yellowish substance. I was not particularly alarmed. I looked up reliable medical studies online and learned that, while this oddity could point to breast cancer, nine times out of ten such discharge was caused by a benign condition. As a journalist, I found this statistic reassuring. In my mind, if there was only a small chance something could occur, it would not occur. Relying on data and evidence over fear and emotion was a hallmark of my work covering health care and politics for *Time* magazine. For years, I had reported on other people's problems—their lack of health insurance, their status as victims of careless government policy, the costs they incurred by getting sick. My work could have tuned me in to the possibility that I might one day find myself as vulnerable as the people I wrote about. Instead, while I felt compassion for those I interviewed, their experiences seemed wholly separate from mine. I felt invincible.

I had never been the type of person who ran to the doctor with every odd rash or small pain. I was a health-care reporter without a general practitioner. I had never broken a bone or confronted any serious health problem. The only discernible cases of cancer in my family were diagnosed in my grandfathers, who had both died in old age. One had smoked heavily and died of lung cancer. The other died of metastatic prostate cancer. In my family, we took pride in not getting sick—as if a genetic predisposition for good health was something we had achieved.

To me, the goo oozing from my nipple was annoying but nothing to panic about. I was only thirty-five, after all. Collin felt differently. Both his maternal grandmother and an aunt had died young of breast cancer. He urged me to get checked out. I promised I would.

First, I had to get through my latest project. That summer and fall, I was working on a story about epileptic children whose brains were ravaged by uncontrollable seizures. Interviewing the parents of these children left me gutted most nights, their suffering shooting through my bones. After long days working on the piece, I felt profoundly sad for the families and worried that I would not be able to convey their stories with enough honesty and empathy. I was stressed, exhausted and, I told myself, too busy to go to the doctor.

But after my piece was finally published in October 2014, I booked an appointment with an internist at the University of California, Los Angeles. During the exam, I brought up the nipple discharge, which had recently turned a darker brown. The doctor's eyes widened. She assured me that it was probably nothing but referred me to a breast surgeon, who ordered an ultrasound. A few days later, a technician chatted with me as she moved a small wand across the front of my left breast. I was relaxed during the exam, but when she left the room, I felt a wave of terror. Was it possible I actually had breast cancer? When the technician returned, she said everything looked good but that I should call the office for the final results in a few days. I felt relieved, but I realized that my body had grown stiff and my jaw ached from clenching my teeth.

When I called a few days later, a receptionist said the doctor who reviewed my ultrasound images had ordered a mammogram. When the plates of the machine squeezed my breast into a flat shape, brown goo oozed from my nipple. I wiped it off with a paper towel before the technician noticed, got dressed and returned to the waiting room, where I scrolled through Twitter on my phone. After a few minutes, a nurse called my name. Instead of telling me I could leave, she led me through the waiting-room door and ushered me into a darkened room with glowing computer screens. As my eyes adjusted to the low light, a radiologist directed me to a screen displaying my mammogram. See there, she said. Squinting, I saw that the radiologist was pointing to some small bright white dots. These are calcifications, the radiologist explained, which can be a sign of breast cancer. We'd like to do a biopsy, she said.

"Is this really necessary?" I asked. As a health-care reporter, I was suspicious. In the United States, unnecessary medical tests waste billions of dollars per year and are often ordered more because of billing and liability concerns than need. Yes, it is necessary, the radiologist said. We walked out of the dark room and into another that had an exam table with a hole in it. I was to lie facedown on the table, depositing my breast into the opening. A different radiologist used a handheld device to shoot a long wide needle into my breast and extract a sample of tissue that would be examined under a microscope. Before he started, the doctor warned me not to be alarmed by the gunshot-like sound of the needle shooting into my flesh. Afterward, the nurse handed me two Tylenol, a cup of water and a small disposable ice pack that I shoved inside my bra on the way out.

I was alone, working in my home office the next day, when a nurse called. I scribbled on a piece of paper as she talked. *DCIS. Stage 0 breast cancer. Schedule appt. with a surgeon.* "A surgeon? Are you serious?" I asked. Yes, she said. Call this number.

I had reported on breast cancer enough to be acquainted with DCIS. A few days later, unnerved by the idea of breast surgery but still not worried my life was in danger, I met with the breast surgeon. Collin came with me and we listened as the doctor prescribed a mastectomy. The DCIS was too widespread for a lumpectomy, she told us, and it had to come out. I burst into tears when she explained that, after reconstructive surgery, I could get a tattoo of a nipple to replace my original one. She gently handed me a box of tissues and said she wanted to order an MRI, just to confirm the mammogram results that showed there was no invasive cancer present. My suspicion about unnecessary testing evaporated. "How soon can I get the MRI?" I asked. It would be at least a week. In the meantime, Collin and I drove to San Francisco to spend Thanksgiving with my in-laws. As we zoomed along Interstate 5, I stared out at the fields of the Central Valley and wondered if I was at the beginning of my breast cancer experience or near the end of it.

Collin had recently given Evie a pale pink Fuji Instax camera, and she had slipped it in her suitcase before the trip. As we lounged in the living room after our Thanksgiving feast, Evie ran into the room with the camera. She wanted to take pictures of everyone. "My whole family!" She shot a photo of Collin smiling wide, my mother-in-law with her beloved golden retriever, Rusty, and my father-in-law sitting at the dining-room table. When she faced the camera toward me, my stomach began to turn.

I smiled. The flash went off. Evie snatched the photo as it glided out of the slot at the top of the camera. She laid the picture on the table and climbed onto my lap. We watched an outline of my head and shoulders begin to appear. As the image became clearer, my sick feeling intensified. I had a vision of this photograph at the bottom of a shoebox Evie would keep on a shelf in her closet, part of a collection of mementos she would look at alone in quiet moments. I imagined her staring at this tiny photograph years later, trying to remember what it felt like to have a mother.

The MRI machine, located in a trailer annex near Kaiser Permanente's hospital on Sunset Boulevard, had two breast holes. I wore earplugs as the enormous magnetic machine whirred. The results came a few days later. They showed that, in addition to a large cluster of DCIS in my left breast, there were two spots nearby that also appeared abnormal. I went back in for another biopsy of those areas, which doctors suspected might be invasive breast cancer. This time, I lay on a table faceup, my left arm extended above my head. The gunshot machine rang out more than ten times as a radiologist took tissue samples of the abnormal spots and two lymph nodes that also looked suspicious. She used an ultrasound wand to guide the needle. I stared silently at the screen as she worked and could see several dark blobs and the biopsy needle extending into them in real time. "You are so brave," the doctor said, a confusing description for the resignation that had settled over my brain.

My surgeon called the next day to report on the blobs. "Unfortunately, it looks like you do have invasive cancer," she said. And just like that, my fear about a nipple tattoo was a distant memory, crowded out by an urgent willingness to do whatever it took to stay alive. "I'd like to order a bone scan and a CT of your abdomen," said my doctor. These additional imaging tests would tell us if the cancer had metastasized and formed tumors elsewhere in my body—stage IV, the point of no return. Doctors do not consider stage IV breast cancer curable. Patients typically undergo continuous treatment until they die.

It was unlikely that I had metastatic disease, but this fact provided no comfort. My short experience as a cancer patient had already taught me that, however unlikely an outcome might be, it was possible to land in that small pool of people who got the very worst news. Somebody had to. Collin and I went to the movies the next day, trying unsuccessfully to focus on something other than the specter of death that hovered above us. Toward the end of the movie, my phone vibrated. I ran out of the theater and heard a scheduler say there was a cancellation in the bone-scan clinic. Could I be there in an hour?

The bone scan complete, a few days later I entered a cold room at Kaiser for a second body scan, a CT. A nurse handed me a hospital gown and a bottle of barium sulfate to drink. The barium, which tasted chalky, had to move through my stomach and intestines for about an hour before the scan. It would coat my digestive tract, allowing a doctor to better map my insides. As I waited for the time to pass, I shivered in the gown and flipped through a hospital pamphlet, the only reading material in the room. Next to the seating area, there were a few empty bays where patients on stretchers could wait for various scans and tests. Before long, an orderly wheeled in an old woman on a gurney. The woman was writhing and moaning loudly. "I'm in pain!" she cried. As the orderly and a nurse tried to calm the woman, I wanted to scream back at her: *You are old. You got to be old. You had a long life.*

I was finally sent to the CT scan suite, where I climbed onto a table and the technician started an IV in my arm through which he would administer a second contrast agent. My eyes welled up with tears. "Don't worry. It won't hurt. We're just going to take some pictures," he told me, putting his hand on my shoulder. I nodded. I was not worried the scan would hurt. I was worried it would prove that I was dying.

I am the type who can go years, even a decade, without shedding a tear. But in the brief period between undergoing the bone and CT scans and learning the results, I wept every day and every night, pleading aloud with a God I didn't believe in to please not let me die. I cried alone and in the presence of my husband, although he never joined me. Later, he told me he had cried only at work. I never broke down in front of Evie, but I stared at her every time we were in a room together, terrified she would sense my distress.

One night, I watched Evie assemble a puzzle on the floor. An image of dancing ballerinas appeared as she put the pieces into place. My father called from New York to ask for an update. Did you get the results? Has it spread? My husband, Collin, stepped into the dark, cold December air to talk to him. For hours that night, I lay on Evie's floor, my hand resting on her back at it rose and fell with her breath. After I peeled myself off the floor, I moved to my bed, where I stayed up until the darkness turned to dawn, the glow of my laptop illuminating my face as I typed various search terms. *Young women, breast cancer, death rate. Breast cancer, young, life expectancy. Breast cancer, lymph nodes, mortality.*

When I talked to my father the next day, I said that if I could just survive until Evie's first day of kindergarten, this whole ordeal might be slightly less horrifying. Let's focus on that, then, he said. Let's focus on keeping you alive for two more years. By the age of five, Evie would be able to form durable memories. She would know me even if she lost me, although I wasn't sure if this was better.

The most logical thing I could think to do was to make lists. A list of things Collin would need to know about girls: ear piercing, how to braid, tampons, the importance of a really great prom dress. A list of books, definitely some Roald Dahl and George Orwell. I should get some big plastic bins, I thought. I could buy all the items Evie would need up until about age twelve, organize and label them, maybe put them in a storage space. I made a list of items: shoes and clothes in every size, a variety of lunch boxes and backpacks, school supplies. Fewer things for Collin to worry about. And my girlfriends. I made a list of the best ones and wrote them letters in my head, asking them to be surrogate mothers to Evie. Why had I never gotten around to buying life insurance?

I made my parents promise to help raise and support Evie financially—as if they would have done anything else. I made my brother promise to take care of Evie if the two of them were ever the only ones left. Over beers one night, I tried to make Collin promise to get remarried, but he wouldn't. Promise me, I said. Maybe after Evie is grown up, he said.

"I feel like the worst-case scenario keeps happening," I said to my surgeon when Collin and I met her in an exam room to discuss the CT and bone-scan results. "I know," she said, "but I think that's stopping now. I don't think there is any more bad news." The scans were clear. I did not have any other cancerous masses. "This is treatable, okay? People can survive this," the doctor said. Still, I needed to act quickly. The type of breast cancer I had was particularly fast-growing. "What's the most aggressive thing I can do?" I asked.

Before my mastectomy, I would need eighteen weeks of chemotherapy, she told me. Although my scans did not show cancer anywhere other than in my breast and armpit lymph nodes, there was a good chance that malignant cells were coursing through my bloodstream and establishing microscopic tumors elsewhere in my body. The chemotherapy and another type of intravenous drug therapy designed to attack my type of breast cancer would hopefully kill off malignant cells no one could see. After that treatment, I would undergo surgery, five weeks of daily radiation sessions and another six months of drug infusions. Our California dreamland would be a staging ground for more than a year of cancer treatment.

There is no way to know when I would have found my breast cancer if not for the strange goo that started leaking from my nipple in the summer of 2014. If it had remained undetected until I had a routine mammogram at forty—if I'd even survived that long—it's very likely the disease would have killed me. There's also, of course, no way to know exactly how long the cancer had been growing by the time it was found. For reasons not yet understood, a healthy cell inside one of my milk ducts mutated, kicking off a rapid cycle of uncontrolled division. Eventually, there was a colony of cancer. Billions of cells divide every day in a human body, making it truly miraculous that mistakes don't slip by our immune systems more frequently. Oncologists who have reviewed my case estimate that the first cell likely went rogue six months to a few years before I knew it. By the time I was diagnosed, hundreds of millions of breast cancer cells were thriving inside me.

Some version of my tumors may have been growing when we made our cross-country move. My breast cancer may have been present during my seemingly uneventful pregnancy, dividing rapidly in the upper outer quadrant of my left breast while, just below, another set of cells was replicating and forming new life. My cancer may have been there when I was most in tune with my breasts, as a new mother, cradling my baby daughter's head in the crook of my arm while she drank my milk.

At the time, finding out exactly why I had developed breast cancer seemed like a luxury. My life was at risk and saving it was my only priority. But in the years since, I have searched my memory, wondering if I did something to cause my disease. I drank alcohol, which increases overall breast cancer risk, but the link is weak or nonexistent for women with the specific type of breast cancer I had. I smoked cigarettes for more than a decade, and studies have found smokers have an increased risk of breast cancer, although the increase is modest. Obesity increases breast cancer risk, but I was thin. I became pregnant for the first time after the age of thirty, which studies have shown increases the risk of breast cancer, but so did many women who never developed the disease. I wondered if I had been exposed to some breast cancer–causing pesticide while growing up on a farm. I wondered if my mother had taken some drug, legal or otherwise, during her pregnancy that could have put me at risk.

Researching my past and perusing the medical literature, I could find no clear reason why I had developed breast cancer at thirty-five. "Cancer happens, even in young people, but there's nothing identifiable that caused yours," my oncologist told me. "There could have been some environmental exposure that we're not aware of." What that exposure might have been, my doctor could not say. Indeed, despite the billions of dollars spent on breast cancer research, scientists know relatively little about what triggers the disease. And large-scale efforts to find out have mostly come up empty.

Lorraine Pace had a hunch. After she was diagnosed with breast cancer in 1992, Pace realized that some of her neighbors in West Islip, New York, a small hamlet on Long Island, were in the same boat. First it was one neighbor, then another, until Pace counted twenty women living near her who had also been diagnosed with breast cancer. Pace got a map, spread it across her dining-room table and started plotting locations, marking households yellow if a woman living there had been diagnosed with breast cancer. In the months that followed, Pace's map got more and more yellow and she began to wonder if there was a connection between all of these cases, some outside factor that was making women sick. Could there be a pollutant in the air or water that was causing breast cells to mutate? Before long, Pace's hunch—that something in the environment had given her cancer—became an obsession. She started a nonprofit advocacy group to raise awareness and mailed questionnaires to residents in West Islip, eventually collecting thousands of responses. A New York State study released in 1990 had found higher concentrations of breast cancer in several Long Island communities, but health officials said the finding could not be tied to any known environmental factor. Pace traveled to Atlanta the year of her diagnosis to ask the Centers for Disease Control and Prevention to reevaluate the evidence.

In its report, the CDC concluded that Long Island's relatively high breast cancer rate was a result of demographics, not the environment. Women who are Jewish, have few or no children or become pregnant for the first time later in life are at higher risk for breast cancer, and this described enough of the female population on Long Island to skew the numbers. (A hereditary genetic mutation that can cause breast cancer is more common in women of Ashkenazi Jewish decent than in the general U.S. population. Pregnancy alters the amount of estrogen a woman is exposed to in her lifetime, which can affect her risk of breast cancer.) Likewise, women who are upper-income and white tend to have more access to mammogram screening and, therefore, higher rates of detected breast cancer. But Pace and other breast cancer advocates on Long Island, who had been galvanized by worrisome stories of a possible breast cancer cluster, were not convinced. Their advocacy caught the notice of a congressman and a senator, who lobbied their colleagues on Capitol Hill to set aside federal money to study the issue. In 1993, Congress passed a law ordering the National Cancer Institute to do just that.

Right around the same time, another state report was published that convinced Pace and her fellow advocates that they were onto something. The health department said that post-menopausal women who lived near chemical plants on Long Island were more likely than other women to develop breast cancer. The report, while worrisome, did not establish a definitive link, but top scientists said it warranted more study. The NCI, Pace hoped, would sort everything out. It would spend thirty million dollars over the next ten years to study breast cancer in West Islip and the surrounding communities, a hefty investment that reinforced the suspicion that something was seriously wrong on Long Island.

NCI researchers set to work gathering data from about three thousand women on the island, half of whom had been diagnosed with breast cancer and half of whom had not. Every woman included in the study provided blood and urine that was tested for traces of dangerous chemicals. Researchers collected samples of tap water and dirt from study participants' yards and used high-powered vacuums to gather dust particles from their carpets. The leader of the effort was a scientist named Marilie D. Gammon of the Columbia University School of Public Health and later the University of North Carolina, Chapel Hill. Like Diana Petitti, Gammon was an epidemiologist who specialized in the population-based study of diseases. In her application to work on the project, Gammon had noted what the CDC had also noticed. Contrary to claims that Long Island's breast cancer rate was inexplicably high, Gammon wrote that the rate was actually pretty average considering the specific demographics of the population. Such a fact might have stopped a researcher from researching, but the Long Island project had been mandated by federal law and so Gammon carried on.

When the NCI finally reported the results of its study to Congress in 2004, Pace and her allies in the breast cancer advocacy community on Long Island were crestfallen. The study had failed to identify any new information that explained why women on Long Island developed breast cancer. The pesticide DDT and electromagnetic fields, two suspected cancer-causing factors that Gammon and her colleagues had studied, did not appear to increase the risk. Another class of chemicals might slightly increase risk, they said, but more study was needed to confirm it. (Subsequent study showed no link.) Researchers wrote that the scientific investigations funded by the NCI project "confirmed the presence of established risk factors for breast cancer among women on Long Island," including Jewish ancestry and childbearing trends. In the end, the project helped launch new cancer registries and backed efforts to increase mammogram screening rates, but it provided no new knowledge about what caused breast cancer or whether the disease was related to water, soil, air or pollutants.

A 2015 study published in the journal *Science* found that two-thirds of the mutations that could potentially give rise to cancers were random and unrelated to hereditary or known environmental or behavioral factors. In a follow-up study published in 2017, the authors parsed the data further and found that fewer than 15 percent of the mutations in breast cancer cases were driven by environmental factors. This does not mean that the environment does not play a role in breast cancer, only that no reliable links have been definitively proven. Historically, Japanese women who migrated to the United States had a significantly higher rate of breast cancer than women in Japan, suggesting environmental or dietary influences might be in play. Breast cancer rates in Japan have steadily increased as more members of its population have adopted a lifestyle and diet influenced by the West.

Scientists have pointed to foods such as green tea and soy as agents that might protect some women from developing breast cancer in Asia, where the rate of the disease is still lower than it is in Europe and the United States, but no large-scale study has ever proved that a particular food can stave off breast cancer. (Claims that sugary food causes or exacerbates breast cancer are not supported by science.) In the absence of more knowledge and information about what causes breast cancer, the authors of the *Science* articles attributed mutations, the most important driver of the disease, to "bad luck." This is understandably maddening, and we are unlikely to know more anytime soon. Compared with the research funds and energy spent on breast cancer treatment, investigation into causes of the disease is anemic at best.

Critics of public- and private-sector research priorities often say no one studies breast cancer causes or prevention because there is no profit in it. For pharmaceutical companies, they say, there is no financial upside to learning why women get breast cancer. But this is not true for the federal government. The United States, through public-insurance programs, spends billions of dollars treating breast cancer patients every year. Finding causes of breast cancer so it can be prevented has a huge upside. And yet, since the Long Island episode, the federal government has launched no full-scale effort to understand what causes breast cancer beyond known risk factors such as obesity, alcohol and having fewer children or having them late in life. (A 2013 report commissioned by the U.S. Congress called for an increase and realignment of research into breast cancer and the environment.) These known risk factors matter—a lot. The chances an American woman will get breast cancer is one in eight. But for a slender woman with no previous cancers and no family breast cancer history who started her period after age eleven, had a baby before age thirty, and entered menopause before the age of fifty-two, the chances are significantly lower. The fact that the women who fit this description do sometimes get breast cancer is proof, though, that we do not yet understand all the factors that contribute to the disease.

There are murmurs in the popular and scientific press nearly every week about small studies examining or indicating possible links between breast cancer and various outside factors. The week that I wrote this paragraph, newspapers and websites around the world reported on a new analysis showing that women who regularly eat processed meat have a 9 percent greater than average risk of developing breast cancer. Read the paper closely, however, and it's clear that bacon doesn't necessarily give women breast cancer. The analysis said that consumption of processed meat in postmenopausal women is associated with an increased risk. Over a lifetime, a 9 percent increase amounts to less than one extra case of breast cancer per one hundred women. There is no clear cause and effect. Even studies that try to control for outside factors are typically conducted on a small scale and within a scope unlikely to produce definitive answers to why some women develop breast cancer and others do not. In a study launched in 2013, for example, researchers are administering a chemical found in olive oil to one hundred women with dense breasts (a risk factor for breast cancer) in an attempt to gauge whether it reduces their breast density.

There is one thing, however, that scientists know causes breast cancer: certain genes. Doctors and women have long known that the disease sometimes runs in families, but for decades no one knew why. The answer came in 1990, when a team of researchers at the University of California, Berkeley, led by a scientist named Mary-Claire King, identified a gene called *BRCA1*. King's work showed that women who had a mutated version of *BRCA1* had a much higher rate of developing breast cancer than the general population. A father or mother whose DNA contained the faulty gene had a 50 percent chance of passing the gene to offspring. King's discovery set off a race to find the location of *BRCA1* in the human genome and clone it, which would enable scientists to develop a test to identify carriers.

Researchers in King's lab at Berkeley worked feverishly to map, sequence and clone *BRCA1*, but they lost the race in 1994 to a team led by scientist Mark Skolnick in conjunction with a private company called Myriad Genetics. Soon, Myriad scientists had located and cloned a second breast cancer gene, *BRCA2*. Taken together, *BRCA1* and *BRCA2* cause fewer than 10 percent of all cases of breast cancer. But for women who have mutations of one or both genes, finding out that they are significantly more likely than average to develop breast cancer can be life-changing. A *BRCA* carrier's risk of developing breast cancer varies by person, but it is high enough that a growing number of women who learn they have it undergo preventive double mastectomies.

Myriad Genetics wanted to corner the market on breast cancer genetic testing and tried to patent *BRCA1* and *BRCA2*. But in 2013, the U.S. Supreme Court ruled that human genes could not be protected by patent law. Today, several companies, in addition to Myriad, offer *BRCA* tests, which have helped reduce the cost of basic screening to a few hundred dollars. After *BRCA* tests came on the market, some public-health officials worried that women without a family history of breast cancer would demand to be tested, opening a can of worms that could lead to widespread unnecessary testing and associated costs. I wrote a story for *Time* about these fears. (A study by two Harvard researchers found that the rate of *BRCA* testing jumped by more than 50 percent nationwide in the two weeks after actress Angelina Jolie published a 2013 op-ed in the *New York Times* about her decision to have a preventive double mastectomy after testing positive for *BRCA1*.) But in recent years, many have been calling for an expansion of testing beyond just women with breast cancer diagnoses and a family history of the disease. In early 2019, the American Society of Breast Surgeons issued a recommendation to offer genetic testing to all women diagnosed with breast cancer. For her part, King says she believes all young women should be screened for breast cancer–related genes even if they have never been diagnosed with cancer. "Every breast cancer patient we identify *after* she develops cancer clearly represents a missed opportunity for prevention," King told an audience in 2015. These days, scientists know that in addition to *BRCA1* and *BRCA2*, other genetic mutations can increase the risk of developing breast cancer. The research is ongoing, but it's likely that breast cancer incidence is more closely tied to hereditary factors than scientists previously believed. (After my diagnosis, I was tested for the *BRCA* mutations and several other mutations; all the tests were negative.)

It's difficult to imagine an effort like the Long Island Breast Cancer Study Project happening today. Politicians and Congress do not typically set research priorities for the NCI, for one thing. A cynical read of the Long Island project might conclude that it was more about elected officials wanting to appease and gain the support of their constituents on Long Island than about science. But beyond science and politics, there is another reason Lorraine Pace and other women prevailed in getting the federal government to pay attention to their worries about breast cancer causes in the 1990s. Back then, spending money on breast cancer had become trendy.

The 1991 Tony Awards took place, as they do every year, in a theater in the heart of midtown Manhattan. Julie Andrews and Jeremy Irons hosted the show, joining Broadway actors, dancers, directors, designers and producers to celebrate the year's best plays and musicals. *Miss Saigon,* the year's highest-grossing new production, won three acting awards but lost the award for best musical to *The Will Rogers Follies*. Neil Simon's *Lost in Yonkers* won for best play. *Fiddler on the Roof* won best revival.

What stole the show, though, wasn't a speech, performance or upset win. It was a strange little accessory worn by some presenters and award winners that reminded those watching of the day's most pressing health crisis. An AIDS-focused art collective had come up with the idea, its members cajoling assistants who helped celebrities dress for the show to pin red ribbons to their bosses' lapels or dresses. Irons wore a ribbon. So did Jonathan Pryce, who won the award for best actor in a musical, and Tyne Daly, who presented the award for best original score. Even eleven-year-old Daisy Eagan, who won an acting award for her role in *The Secret Garden,* wore a ribbon. No one publicly mentioned the AIDS crisis that night, but no one needed to. Almost immediately, the red ribbon went from an art project conceived by New York AIDS activists to an international symbol of the movement to raise awareness and find a cure for the disease.

Three thousand miles away, a California housewife named Charlotte Haley had an idea. If a red ribbon could galvanize support for AIDS research, maybe a ribbon could do the same for breast cancer, which killed more Americans and was, like AIDS, a disease plagued by stigma. Haley had never had breast cancer, but her family had been ravaged by it. Her grandmother, sister and daughter had all had the disease, and Haley wanted to do her part to increase funding for research on it, the same way AIDS activists had done. Breast cancer needs its own ribbon, she thought. Haley's favorite color was peach, so she settled on that hue, buying reels of ribbon at a craft store. At home, she and her grandson cut and folded the peach ribbon into loops that they attached to small cards. On the cards Haley typed messages urging people to lobby for an increase in funds for breast cancer research and prevention. Haley gave her ribbon cards to her friends. She dropped off stacks at doctors' offices and handed them to customers at her local grocery store. Then she started to get more ambitious, sending cards to several former First Ladies and a few newspaper columnists, including Liz Smith, a gossip writer in New York. Smith was intrigued by Haley's homegrown effort and published a column about the breast cancer–ribbon cards along with Haley's phone number. Pretty soon, Haley and her husband, Bob, were so besieged with requests for the ribbons that they turned their living room into an assembly line, eventually mailing tens of thousands of ribbons to people around the country and beyond.

Among those who learned of Haley's grassroots activism was Alexandra Penney, the editor of *Self* magazine. *Self* had published its first breast cancer–awareness issue the year before, and for the magazine's 1992 effort Penney wanted to do something higher profile. She enlisted the help of makeup magnate Evelyn Lauder and called Haley, hoping the California housewife would jump at the chance to publicize her peach-ribbon campaign on the cover of a national magazine.

To Penney's surprise, Haley said no. She didn't like the idea of teaming up with a corporation like Estée Lauder. Her ribbons were effective, she thought, because they were homemade and uncorrupted by business interests. But Penney didn't give up. She consulted with lawyers who said *Self* could do its own breast cancer–ribbon campaign as long as they used a different color than Haley's peach. Penney settled on pink, debuting the ribbon on the cover of *Self* in October 1992. Estée Lauder handed out more than one million pink ribbons at its makeup counters, and before long *Self*'s pink ribbons overtook the peach ones cut in Haley's living room.

The national movement around the pink ribbon succeeded in part because it hit a generational sweet spot at the right moment. Many of the most powerful breast cancer–advocacy groups in America were founded by the same capable generation of women who marched for *Roe v. Wade* and were entering their forties and fifties in the 1990s. These groups raised money that poured into efforts to support breast cancer patients, fund scientific research and raise awareness about the disease. The National Breast Cancer Coalition, a powerful lobbying group dedicated to the disease, was founded in 1991, and the following year it persuaded the Department of Defense to start a breast cancer research program with twenty-five million dollars in funding. (In the years since, the DoD has spent nearly three billion dollars on the program.) The Avon Foundation launched its Breast Cancer Crusade initiative in 1992. The Breast Cancer Research Foundation, a powerhouse nonprofit that raised some seventy million dollars in 2017, was founded by makeup heir Evelyn Lauder in 1993. That same year, a cover of the *New York Times Magazine* featured a photograph of a woman revealing a mastectomy scar across the right side of her chest. It didn't hurt that women who wanted to increase awareness of breast cancer and money spent on the disease had an ally in the White House. President Bill Clinton's mother had been diagnosed with breast cancer in 1990 and died of the disease in 1994.

Partnerships between breast cancer charities and big business have only increased since the 1990s. Charities like the arrangements because they bring attention to breast cancer and raise money. Corporations jump to partner with breast cancer charities because doing so improves their images and, although difficult to measure precisely, their bottom lines. Breast cancer itself became a brand in the 1990s, and the brand endures. What's a better sales pitch than promising consumers that the products they buy will help save the lives of America's mothers, daughters and sisters?

It's an old story by now, but the swath of companies that have marketed items with a breast cancer tie-in has expanded to such an extent that it seems there is not a consumer-product category in America that is pink-ribbon-free. In any given year, you can support the cause of breast cancer by purchasing windshield wipers, blenders, downhill skis, sneakers, tool sets, perfume, water bottles, lipstick, sunglasses, golf balls, jeans, hamburgers, yogurt, chicken sausage, wine or beer. You could fly in a plane painted with a pink ribbon (Delta), eat a bagel twisted into the shape of a ribbon (Panera Bread) or watch television in the name of breast cancer (Lifetime). You could even watch porn to support the cause (Pornhub). In her seminal 2001 essay "Welcome to Cancerland," journalist Barbara Ehrenreich noted that in the 1990s, breast cancer "blossomed from wallflower to the most popular girl at the corporate charity prom."

Some companies that have tied their brands to breast cancer donate a percentage of the sales of each product to various pink-themed charities. Others cap donations once they reach a predetermined amount. Some companies do not send any money to breast cancer charities, the theory being that simply selling an item adorned with a pink ribbon helps support women.

Over a three-day period in October 2011, construction workers in London hoisted a metal frame up the side of a building owned by British broadcaster ITV. Dangling from cables high above the city, the workers had a bird's-eye view of the river Thames. The frame they were positioning measured about one hundred feet wide and extended from the roof across five stories of the building. When it was secured, the workers slowly and methodically stretched nearly five hundred square yards of fabric across the frame.

The world's largest bra, a mega-size 34B, was unveiled on British television on October 28, 2011, earning a Guinness World Record. Reckitt Benckiser, a maker of health products and other consumer goods, paid for the installation, hoping to tie the breast cancer–awareness effort into a public-relations campaign for a stain remover. (In 2012 a casino paid about six thousand dollars for the garment on eBay, money that went to a breast cancer charity.) In addition to the largest-bra record, breast cancer–awareness campaigns hold at least ten other Guinness titles. In 2014, the world's largest T-shirt, a two-hundred-by-three-hundred-foot pink top, was unveiled in Brazil. Race-car drivers in Riyadh, Saudi Arabia, made the world's largest ribbon skid mark in 2016, expertly drifting their pink smoke-spewing cars in loops. The world's longest tutu was wrapped around a bus in Manchester, England, the same year. In 2018, 1,956 women in Pune, India, painted their nails simultaneously to set a record and raise breast cancer awareness. More than twenty-five hundred people jumped naked into the Irish Sea a few months later to do the same. There are breast cancer awareness–related records for the longest bra chain (Australia, 2009), the most landmarks lit up in pink in a single day (worldwide, 2010) and the most mammograms performed by a single provider in twenty-four hours (United States, 2013).

For all the awareness brought about by corporate campaigns and breast cancer charities, most women I interviewed for this book—breast cancer patients and nonpatients alike—know very little about breast cancer itself. They don't know that it comes in different varieties, that a significant share of women diagnosed with invasive breast cancer will eventually see their cancer recur, that treatment has gotten far better and easier to endure, that mammogram screening is deeply flawed. It's as if public understanding of breast cancer was frozen in time in the 1990s.

The pinking of America has divided breast cancer–advocacy groups. On one side are organizations worried that all this "pink-washing" has squashed an attempt to further educate the public and obscured the needs of breast cancer patients whose disease fails to respond to treatment and who, therefore, do not fit the hopeful message of "survivorship" inherent in campaigns backed by corporate interests. Pink is a cheerful, sexy color, they say, inappropriate for a deadly disease. On the other side are groups that say so much pink is the point, a color that unifies women and has helped breast cancer become part of the mainstream conversation in America. And no group has stuck by pink more fervently than the Susan G. Komen foundation.

4

pink vibes

Shirley Temple Black first felt the lump in her breast in September 1972. Her doctor recommended she get a mammogram and then a biopsy. The latter would have to wait. Black, then forty-four, had business to take care of. While Black had tap-danced and sung her way to stardom in dozens of 1930s movies, her life as an adult had taken a serious turn. She had run for Congress in 1967 and lost, but in 1969 President Richard Nixon appointed her as a U.S. delegate to the United Nations. By the fall of 1972, Black was working for the president's council on environmental policy, and after her mammogram she was sent to attend a meeting in Moscow.

When she returned home, Black prepared for her biopsy and prepared for the worst. She read everything she could find about breast cancer. She scanned newspapers and asked her brother, who worked as a hospital administrator, to send her scientific articles. As she read, Black became increasingly alarmed to learn that the biopsy she had scheduled for November 1972 might lead to immediate amputation of her breast. In the early 1970s, breast biopsies were done under general anesthesia in the operating room. The suspicious tissue was examined under a microscope, and if it was found to be malignant, a mastectomy was performed while the patient was still asleep. Women would wake up from surgery, check the clock on the wall and feel their chests. If only an hour had passed, the mass was benign. If the surgery took longer, a mastectomy had been performed. Right around the time Black found her lump, some doctors and breast cancer patients had begun to criticize this so-called one-step procedure, which excluded women from the decision-making process of whether to have their breasts removed.

As Black learned more about the issue, she began to feel something more than alarm. She felt a sense of activism bubbling up inside her. Women need to know they have choices, she thought. Black underwent a surgical biopsy but forbade her surgeon from removing her breast. The lump was determined to be malignant and she later had a mastectomy. But unlike a lot of women diagnosed with breast cancer back then, Black did not want to hide away as she recovered and came to grips with her new, deformed body. "I wonder if I can turn this adversity into some help for my sisters," she told her family. Black invited a friendly local reporter to her bedside in Palo Alto and allowed him to photograph and interview her in the hospital; he sent a story about her ordeal out over the newswires. Some fifty thousand letters of support from readers across the country poured in and Black decided to take her publicity campaign even further, writing an essay for the February 1973 issue of *McCall's* magazine titled "Don't Sit Home and Be Afraid."

In the 1950s, a society editor at the *New York Times* had

reportedly once refused to print a notice for a breast cancer support group, deciding the word *breast* was taboo. In the decades that followed, the American Cancer Society urged women to ignore the stigma around breast cancer, monitor their breast health and alert their doctors about lumps and pains. Black's decision to go public with her cancer diagnosis and surgery sent a new message. The disease was nothing to be embarrassed about. In the context of the women's health movement, which began in the 1960s, Black's public declaration struck a chord. The U.S. Supreme Court had decided *Roe v. Wade* the month before her *McCall's* essay was published, and women were already marching in the streets for more control over their bodies and health decisions. The following year, 1974, First Lady Betty Ford and Happy Rockefeller, the wife of the vice president, also went public with their breast cancer diagnoses. In 1975, breast cancer patient and activist Rose Kushner published a groundbreaking book about her experience. Like Black, she had been advised to have a one-step procedure but refused. In 1976, NBC News reporter Betty Rollin published a memoir about her breast cancer diagnosis and treatment. Both books were top sellers. Breast cancer was no longer a shameful malady quietly and exclusively managed by doctors, who were mostly men. It was out in the open, and women were writing about it and influencing decisions about the disease in the operating room and the government.

But widespread change takes time. After Nancy Brinker's sister, Suzy, was diagnosed with breast cancer in 1977, a surgeon in Suzy's hometown of Peoria, Illinois, removed her tumor and said she was "cured." Relieved to have her breast cancer experience behind her, Suzy didn't ask questions. But her cancer recurred just six months later, appearing in her lymph nodes and right lung. She underwent radiation treatment and a second surgery at the Mayo Clinic in Rochester, Minnesota, but her cancer continued to progress. Brinker, who lived in Texas, had tried to persuade her sister to see an oncologist at M. D. Anderson in Houston. Finally, Suzy agreed. The Houston oncologist she met with prescribed chemotherapy and kept the mother of two alive for three more years. She died in 1980 at thirty-six.

Her sister, meanwhile, married a wealthy Dallas businessman and carried on raising her only child. But the death of Suzy, her only sibling, was devastating and she did not want it to go unmarked. So, two years after Suzy's death, Brinker started the Susan G. Komen Breast Cancer Foundation in her sister's name and planned a fund-raiser luncheon at her husband's polo club. Through her husband's business connections, Brinker got in touch with Betty Ford and invited her to come. She did. Brinker's first fund-raiser was a success, allowing Brinker to award thirty thousand dollars in grants to breast cancer researchers at M. D. Anderson and Baylor University, so she kept going. In 1983, she combined her by-then-annual Dallas fund-raiser with a 5K "fun run" that drew some eight hundred participants. The next year, Brinker herself was diagnosed with breast cancer. Like her sister, Brinker underwent surgery and chemotherapy. Unlike her sister, Brinker survived.

By 1989, Brinker had opened a second chapter of the foundation in San Francisco and started a toll-free breast cancer help line. In 1991, the foundation gave out one million dollars in grants. By 1995, the foundation staged fun runs, rebranded as the Race for the Cure, in fifty-seven U.S. cities. By 2011, the organization and its affiliates across the country were awarding more than three hundred million dollars a year in grants for research, education, advocacy and support for breast cancer patients. All told, the organization had raised nearly two billion dollars. Money went to groups promoting breast cancer screening and education but to research too. Early Komen grants helped fund research pioneers such as Mary-Claire King, whose lab first identified the *BRCA* gene, for example.

As Komen and its races grew, so did criticism of its inner workings. Although the group gave away nearly seventy million dollars in research grants in fiscal year 2011, the federal tax return for Komen's headquarters in Dallas showed the organization spent twenty-six million dollars on employee benefits and salaries, including the nearly seven hundred thousand dollars Brinker was paid. That same year, according to its federal tax return, Komen spent twenty million dollars on advertising and promotion and twenty-one million on consultants and other professional services. Brinker flew first class when traveling on behalf of the organization, which had changed its name to Susan G. Komen for the Cure in 2007. The group trademarked the slogan "For the cure" and sent cease-and-desist letters to other breast cancer charities that used the phrase in their marketing. Komen's many corporate partnerships also drew criticism. It seemed there was no company with whom the group would not raise money. In 2010, for example, the group teamed up with KFC on a "Buckets for the Cure" campaign, an odd juxtaposition given that obesity is a major risk factor for breast cancer. Most pointedly, critics noted that Komen's races, with their pink T-shirts and boas, made breast cancer seem like something to celebrate. Patients with incurable metastatic breast cancer, in particular, said they felt left out and left behind by the Komen-led drive to promote survivorship. Of the money Komen spent on programs and services, just 23 percent went to research.

Despite the criticism, Brinker and the Komen foundation persevered. In 2011, the year before everything went wrong, the group was the largest breast cancer charity in America.

Paula Schneider was lying in bed one day, sick from her latest round of chemotherapy, when the mail arrived. Among the bills and credit card offers was an envelope from Avon. The form letter inside explained the company was sponsoring a thirty-nine-mile annual walk to raise money for breast cancer. Schneider had no idea how Avon had gotten her name and address, but the idea intrigued her. Chemo had hit Schneider particularly hard. The week after each infusion, she could barely walk to the bathroom, let alone thirty-nine miles. Still, she thought that signing up for the fund-raising event might motivate her to regain her strength once her treatment was over. Plus, Schneider was the president of a major swimwear company and had a long list of friends and colleagues she could tap for donations. "People feel very sorry for you when you have a bald head," she said.

Schneider had discovered her breast cancer in 2007 while soaping up in the shower. She had run her hand across her armpit and felt a lump. Then forty-nine, Schneider had been getting annual mammograms, which all came back clear, but her mother had been diagnosed with breast cancer years earlier and her maternal grandmother had died of the disease. After some tests, Schneider was diagnosed with triple-negative breast cancer, a particularly aggressive type of the disease. After several months of chemotherapy, she had a double mastectomy followed by thirty-six rounds of daily radiation. She worked a flexible schedule during chemotherapy and took a few weeks off for surgery, but she kept her job at the swimwear company and wore her wig to Portugal for a sales meeting the day after her final radiation session.

Conventional wisdom says that a breast cancer patient who survives for five years after treatment without a recurrence of her disease is cured, but in many cases this milestone is not all that meaningful. Some types of breast cancer can recur decades later. But triple-negative breast cancer, the kind that struck Schneider, is different. When it recurs, it most often comes back within five years. Schneider participated in the Avon walk every year until she passed that critical five-year mark, by which time she had raised some four hundred thousand dollars in donations. "I literally wore out almost all my friends," she said. Feeling good about her contribution and eager to move on with her family and career, Schneider refocused on her work in the corporate world, where she had earned a reputation as a highly capable leader and problem solver.

In early 2015, Schneider became head of a company drowning in problems. American Apparel, the maker of so many hoodies and T-shirts, was in freefall before Schneider was hired to be its CEO. The company's founder, Dov Charney, was under an internal investigation, accused of sexually harassing employees. (Several lawsuits outlining these allegations were settled out of court.) American Apparel's financial health wasn't any better. Loaded with debt and facing layoffs and store closures, the company brought in Schneider to stop the bleeding. She cut overtime and instituted furloughs, which drew the ire of American Apparel's employees, who saw her as a hatchet woman.

Below Schneider's seventh-floor office in the company's factory in downtown Los Angeles, employees protested in the parking lot every Wednesday. Schneider saw her face on some of the signs, crossed out or above messages such as *A woman that's hurting women. CEO Paula Schneider.* Once, someone made a piñata of Schneider that was bashed to bits. It got so bad she hired bodyguards to escort her to work.

Schneider told me she is a naturally measured person. She had become even more so in the wake of her breast cancer diagnosis and treatment. "Once you've had a life-threatening experience, most things don't bother you as much," she said. While workers protested her leadership, Schneider helped restructure American Apparel's debt by ushering the firm into bankruptcy proceedings; the company's share price had plummeted from more than fifteen dollars a share in 2007 to less than a dollar. She transitioned American Apparel back into a private company, streamlined operations and outsourced the making of some products to other manufacturers. By the time Schneider resigned in 2016, American Apparel was on a more stable financial footing. When she announced her departure to the company's board of directors, Schneider told me that she had recommended they replace her with another former cancer patient. No one else could bring the kind of cool head needed to keep the company in business, she told the group. In addition to being known as a strong leader and problem solver, Schneider had now earned another moniker—turnaround artist.

In 2017, Schneider was working as CEO of a company that owned the denim manufacturer 7 for All Mankind when she was invited to attend a retail-business conference at the Four Seasons in Miami. She was being honored with a leadership award and was asked to deliver a speech on empowerment. The night before, Schneider sat in her hotel room trying to think of what to say. "It was supposed to relate back to something that had to do with retail and I had nothing," Schneider said. She had never earned a business degree but had managed to excel in the field. But that field had become less exciting to Schneider, with fickle customers turning away from brick-and-mortar stores in favor of online outlets. So when she stepped to the podium to accept her award the next day, Schneider decided to talk about breast cancer. She told the conference crowd she had felt most empowered when she was a patient and her friends and family had rallied around her.

The speech was a hit. As the applause died down, Schneider returned to her seat. A friend sitting at her table had an idea. She had heard that Susan G. Komen for the Cure was looking for a new CEO. Was Schneider interested? "Yes, actually, I am," Schneider told her. Raising hundreds of thousands of dollars for Avon's breast cancer fund had felt good and she was good *at* it. The idea of working for a cause instead of a corporation was so appealing that Schneider resigned from her CEO job the same month she first interviewed with Komen. "I worked out an exit plan not knowing the least bit whether I was going to have this new job or not," Schneider told me. After a long interview process, Komen hired Schneider in the fall of 2017. She seemed to have everything the organization was looking for. She had had a personal breast cancer experience and was an executive with years of corporate leadership experience. But, just as important, she knew something about turning around an organization in crisis.

* * *

Nancy Brinker has always been a devoted Republican. Since the early 1990s, she has donated hundreds of thousands of dollars to the party, its candidates and Republican-aligned political action committees. In addition to writing checks, she served in the George W. Bush administration as U.S. ambassador to Hungary (from 2001 to 2003) and as White House chief of protocol (from 2007 to 2009). The Susan G. Komen breast cancer foundation, however, had always been nonpartisan. Like most large charities, it historically stayed clear of politics.

But in 2011, Komen brought politics into its executive offices, hiring Karen Handel to serve as its senior vice president for public policy. Handel was an outspoken conservative Republican from a conservative state. She had been Georgia's secretary of state and had mounted an unsuccessful run for governor in 2010. Among her campaign pledges was a promise to end state funding for breast cancer–screening programs provided by Planned Parenthood.

At Komen, one of Handel's first projects was to find a way for Komen to do the same. Komen had long given grants to Planned Parenthood, which provides clinical breast exams and mammogram referrals. The money was a small portion of Planned Parenthood's more than one-billion-dollar annual budget, but it helped pay for screening for low-income women. Pro-life activists and some Catholic bishops publicly opposed the grant funding, which put Komen in the line of fire.

In a book she later wrote about the episode, Handel said the idea to end Planned Parenthood funding was about economics, not politics, and was already in the minds of Komen executives when she began working for the organization. It was

Handel who devised the strategy. They would blame Congress. After failing to garner enough votes to end federal funding for Planned Parenthood, Republican members of Congress had recently launched an investigation of the organization's finances. Komen could claim it was against the group's guidelines to donate to any organization under government investigation. Komen would tell Planned Parenthood it was ending its grant funding but keep the decision out of the public eye. In retrospect, the plan reflected neither the deft skills of a politician nor the kind of forethought one might expect from an organization whose primary currency was its visibility.

When the news broke on January 31, 2012, that Komen was pulling its funding, the reaction was swift and fierce. On February 2, twenty-six U.S. senators sent a letter to Komen urging the group to restore the grants. Michael Bloomberg promised to donate two hundred and fifty thousand dollars in matching funds to help make up for the money Planned Parenthood would lose. Online, the criticism was deafening. More than one million negative Tweets were posted in the days after Komen's plans leaked out, many blaming Handel directly. An online petition that urged critics to contact Komen directly generated so many messages that the organization's servers briefly crashed, according to a former employee. Planned Parenthood raised four hundred thousand dollars in online donations in the twenty-four hours after the decision became public. In the three days it took Komen to reverse the decision, it became clear why charities like Komen stayed out of politics. Breast cancer affects women regardless of party, but Komen's decision to wade into a politically charged issue had made the largest breast cancer nonprofit in America its most hated charity, with pro-choice activists and Planned Parenthood supporters calling for Brinker to resign. Instead, she said she was sorry. "We want to apologize to the American public for recent decisions that cast doubt upon our commitment to our mission of saving women's lives," she said in a statement. Handel resigned, claiming she was a scapegoat.

But the damage to Komen's reputation was done. In fiscal year 2012, Komen had $399 million in revenue. The following year, revenue dropped to $325 million, and it went down to $287 million the year after that. Komen executives said the Planned Parenthood episode—which some in the office called "the fiasco"—wasn't solely to blame, pointing to the economic downturn affecting all U.S. charities. But compared with its counterparts, Komen was doing worse. In 2012, Komen was the forty-third-largest charity in the United States. By 2014, it had slipped to sixty-second and it continued to drop down the list.

The problem, to many inside and outside Komen, was that Brinker refused to step down. Even though she had founded the organization and built it into a powerhouse, Brinker would forever be associated with the Planned Parenthood debacle. Some Komen employees and board members privately urged her to step aside, according to several former employees. As long as she remained in a leadership role, they argued, the organization's fund-raising could not recover. (A Komen spokesman said in an e-mail, "It would not be appropriate to speak to what former board and staff members may have been thinking seven years ago.") In 2012, with Brinker's blessing, the Komen board began looking for a new leader, eventually settling on Dr. Judith Salerno, a Harvard-educated physician who had worked at the National Institutes of Health and had been the executive director and chief operating officer of the Institute of Medicine of the National Academies (now known as the National Academy of Medicine). But although Salerno took over Brinker's CEO job in mid-2013, Komen's founder stayed on as the organization's chair of global strategy, a newly created position that allowed Brinker to remain on board.

According to a former Komen insider, Salerno had been reluctant to accept the job. Komen needed a total reboot, and with Brinker still connected to the organization, large-scale change would be hard and Komen would remain a target. Brinker loyalists inside Komen had grown accustomed to the way the group operated. (Under the terms of her separation agreement with Komen, Salerno is not able to comment on her tenure.) Although Komen had its headquarters in Dallas, its largest source of revenue, the annual Race for the Cure, was managed by more than one hundred independently operated affiliates across the country. Affiliates kept 75 percent of the money they raised for programs and services in their communities and sent 25 percent to Dallas. The Planned Parenthood episode had strained relations with several of the outposts' leaders, some of whom said they did not know of the decision to cut grant funding to the organization until they read about it in the press. (When I asked the Komen foundation about this, a spokesperson for the group said affiliates were told about changes to grant requirements that affected funding for Planned Parenthood.) The resulting backlash had so severely damaged affiliates' ability to raise money that at least one threatened to withhold the funds it would normally send to the mother ship.

Komen as a whole was unwieldy and Salerno wanted to streamline it. She cut expenses and urged affiliates to merge.

But Salerno wanted to do more than reorganize Komen and move on from the Planned Parenthood issue. The organization's brand had become a liability instead of a strength. In the wake of the Planned Parenthood mess, criticisms about Komen's mission had grown louder. The organization spent far more money on education and screening than on research, which drew fire. Metastatic breast cancer patients castigated Komen for focusing on women who survived breast cancer. Young women, who had led the charge against Komen through the Planned Parenthood incident, didn't trust the group.

As CEO, Salerno tried to set Komen on a new path. A few months before Salerno's three-year contract ended in 2016, Komen announced an ambitious goal to cut U.S. deaths from breast cancer in half by 2026. The organization would accomplish this, Salerno said, by refocusing its research-funding efforts on women with aggressive types of breast cancer and metastatic disease. Salerno, a public-health expert, said the organization would also spend more time and money addressing the racial disparities in breast cancer. Black women are diagnosed with the disease more often than other races, but they are 40 percent more likely to die of it, due to lack of access to high-quality treatments and other factors. "This constitutes a public health crisis that must be addressed," Salerno said in a press release announcing the new strategic plan and a twenty-seven-million-dollar grant from an outside organization to fund the effort.

A few weeks later, Komen also announced it had a new slogan: "More than pink."

Komen's headquarters are located in a nondescript office park north of downtown Dallas. When I arrived on a warm Friday in August 2018, Margie Cliff met me in the lobby of the building where Komen occupies two floors. It used to lease three but recently consolidated. Cliff has been an executive assistant to all the CEOs who have led the organization during her three years at the company, including Salerno, an interim CEO, and its current leader, Paula Schneider.

Cliff led me through a maze of desks and offices. On one wall, I saw framed oil paintings of Brinker and her sister, Suzy. On others, I saw a series of slogans. WE DON'T RACE TO BE FIRST. WE RACE TO MAKE SURE THAT ONE DAY, THERE WILL BE A LAST. And THIS IS NOT A WORK STATION. THIS IS YOUR TRAINING GROUND FOR EMPOWERING PINK. Meeting rooms were named for breast cancer buzzwords: *Celebrate. Survivor. Empower. Cure.* A larger room had a plaque outside the door that read THIS IS NOT A CONFERENCE ROOM. THIS IS A STADIUM WHERE WE WILL CONQUER CANCER. On a divider separating desks in an open-plan area: THIS IS NOT A CUBICLE. THIS IS A STARTING BLOCK FOR FUTURE RACES, LABORATORIES, AND TREATMENT CENTERS.

Sean Tuffnell, Komen's director of communications, explained that an Ohio-based marketing company had recently given the offices a makeover. "It's called *environmental branding,*" Tuffnell told me. It seemed Komen's effort to rebrand its mission and reputation extended all the way to the offices and desks at its mother ship. As Tuffnell led me on a tour, he pointed out a sea of cubicles. "This is our affiliate network area," Tuffnell said. It was empty. "We have a flexible work environment," Tuffnell said. "People can work at home one day a week, and everybody picks Friday." The efforts to pare down the number of Komen affiliates operating across the country continued even after

Judith Salerno left the organization in 2016. Some chapters closed because they could no longer raise enough money to be sustainable. The total number of affiliates today is about half what it was before the Planned Parenthood episode.

I met Schneider in her corner office. She had been on the job less than a year and, because she'd had no professional experience in fund-raising or public health, her learning curve had been steep and, at times, emotional. Shortly after she was hired, a former oncology nurse on Komen's staff gave Schneider a presentation known around the office as Breast Cancer 101. Over the course of three hours, the nurse provided an overview of the breast cancer diagnosis, treatment and survival landscape. Schneider listened but eventually said the meeting had to end. "I was starting to relive the experience," she told me. Schneider's own breast cancer ordeal was a decade behind her, but she said hearing about everything that could have gone wrong upset her. Likewise, Schneider said that as CEO of Komen, she'd visited some chemotherapy infusion centers early in her tenure but she'd had to stop. "I told my team I didn't want to go anymore," she said. "I know what they look like."

Salerno's initiatives—focus research on the sickest breast cancer patients and try to close the racial-disparity gap—remain in place. "There's been a pretty seismic shift in this organization," Schneider said. Patients with metastatic breast cancer have been among the loudest critics of Komen over the years, accusing the charity of celebrating those who survive breast cancer and ignoring those who do not. Race for the Cure, they point out, implies that breast cancer is curable for everyone, but it's not. Komen's focus on screening mammography had also rankled groups devoted to patients with metastases, many of whom had gotten annual mammograms and still ended up with terminal disease. "They are angry because they've done everything that they were supposed to do—the things that Komen said to do. They did everything right and they still got metastatic breast cancer," Schneider said. As part of its goal to reduce breast cancer deaths by half, Komen has been organizing conferences geared toward metastatic disease and partnering with some stage IV groups that had previously criticized the organization. "I am only here because I care about women dying and figuring out what has to happen to change that," Schneider said. In fiscal year 2018, 70 percent of Komen's research grants went to scientists studying metastatic breast cancer and what happens when breast cancer treatments stop working.

Komen's most recent tax returns show the organization is raising and spending half as much money as it did before the Planned Parenthood incident. Schneider said she hopes to make up the shortfall by increasing corporate partnerships and decreasing Komen's dependence on annual races and walks, which are drawing fewer participants every year. "Every organization has to evolve," she said. "You're either going up or you're going down."

Victoria Wolodzko, Komen's senior vice president of mission, told me she had recently attended an event in Washington, DC, hosted by a metastatic breast cancer group that advocates on behalf of stage IV patients. Wolodzko said her coworkers told her she was "brave" to go to the event, which was organized by the kind of advocates who had been bashing Komen for years. "I represent what's supposed to be an answer and we haven't given it," Wolodzko said. "It's okay to admit we didn't get it right for everybody."

I asked Schneider if her experience at American Apparel had prepared her to take the reins at Komen. "Once you work at an organization that needs some turning around and you do it, you become that person," she said. "Komen is stable. Okay? This is a stable organization. Has it gone down with fund-raising? Yeah. It has gone down. Are there opportunities for us to do things differently? Oh yeah. There are absolutely opportunities."

As Tuffnell led me back to the elevators, we walked past a time line printed on the wall, another element of Komen's new environmental-branding project. There was a picture of Brinker and Komen, smiling wide, and a time-line entry in 1982 for Komen's first fund-raising event. The logo for American Airlines, Komen's first corporate sponsor, was painted on the wall in pink. One entry noted that in 1991, Komen had raised one million dollars in a single year for the first time. Another reminded those walking past that the Race for the Cure became the world's largest registered 5K in 1998. An entry for 2014 reported that that year, Komen's total fund-raising and spending crested two and a half billion dollars. The time line ended there. But next to it, in white type against a bright pink wall, were these words:

> *We tore down the stigma and raised awareness. We made silence uncomfortable and gave survivors a voice. We've spent millions on research and saved millions in the process. And now, we have more to do.*

When you're first diagnosed with cancer, you want answers. First and foremost: Will this kill me? Will I see my kids grow up? Will I lose my hair? Will it hurt? Why did this happen? Good cancer doctors answer this first wave of questions with honesty and compassion. But when it comes to explaining the biology of a malignant tumor and what modern science has devised to beat it back, even the best doctors need more than the truth and a good bedside manner. This is why they tend to talk in analogies.

The surgeon who diagnosed my breast cancer described chemotherapy as "a weed killer" that could hunt down and kill the breast cancer cells that had broken off from my primary tumors and traveled around my body. She said these lonely cells were like the white "fuzzy things" blown off a dandelion. After my mastectomy, a plastic surgeon would place an expander inside my chest to stretch the muscle and skin to make room for an implant. "It's like a spare tire," she said.

But no cancer expert I've met has honed the analogy game better than Dr. Susan Love, author of one of the world's best-selling breast cancer books. Called simply *Dr. Susan Love's Breast Book,* the tome was first published in 1990 and is currently in its sixth edition. More than one million copies have been sold and have made Love a sort of medical celebrity. Full of complex ideas in easy-to-understand plain language along with illustrations, the book has helped demystify not just breast cancer but breasts themselves. The latest edition of the book, published in 2015, opens with an explanation of how breasts form. Stem cells behind a female baby's nipple are, writes Love, "Like those capsules you had as a kid that held collapsed sponge animals. When you added water, a sponge animal appeared." Human hormones are the water that, once added, causes breasts to form and expand.

After flirting with the idea of becoming a nun, Love had trained to be a surgeon when it was rare for women to pursue the specialty. She established and ran UCLA's groundbreaking breast cancer center from 1992 to 1997. There, Love founded the country's second major multidisciplinary breast cancer clinic. (The first was the Dana-Farber Cancer Institute, where Love worked before UCLA.) At Love's UCLA clinic, newly diagnosed breast cancer patients could see their whole team of doctors in a single visit, a patient-friendly model since adopted at many other major cancer centers. Love was obsessive about the needs of her patients. She got tape recorders for the breast clinic and told doctors on staff to record their consultations and give the tapes to patients on their way out so they could have a second, or third, chance to digest information thrown at them in the midst of visits that were emotional and confusing. Doctors resisted the idea until they realized the tapes cut down on follow-up calls.

At UCLA, Love's colleagues either adored or resented her. Some felt she understood the needs of breast cancer patients better than anyone in the field. Others criticized her for promoting her book and giving speeches when she should have been raising money to support UCLA's breast cancer work.

When I met Love for the first time in the San Fernando Valley offices of the nonprofit foundation that bears her name—she retired from UCLA in 1997—she peppered her answers to my questions with analogy after analogy. Some tumors are like a gang that infiltrates a bad neighborhood, she said. (You can banish the gang, but if you don't change the environment, the danger isn't over.) Others are more akin to terrorists. (You kill one, and another pops up to take its place.) Mammograms are like security guards that patrol a spot every two hours. (They catch some criminals, but only the slow ones.) Ductal systems are like the veins in blue cheese. (They run through the breast in unpredictable sprawling patterns, not straight lines.) Breast cancer cells that metastasize are Cuban boat people. (They escaped.)

As the influence of Komen and other legacy breast cancer charities has waned in recent years, Love's foundation is among the small, more focused nonprofits that have sprung up to fill the void. Love's title at her foundation is "chief visionary officer," and the research she supports often lies outside the bounds of traditional breast cancer science. One foundation project will study the breast's microbiome, for instance. Another Love-guided project is mapping milk ducts inside the female human breast, an area of the body that has been profoundly understudied. The last milk ducts to be activated when a woman breastfeeds are those in the upper outer quadrant of the breast, the same spot where breast cancer most frequently occurs. No one knows if there is a connection. Likewise, after a woman stops breastfeeding, her milk ducts go through a process called *involution,* shrinking and returning to a dormant state. No one knows for sure how this might relate to breast cancer that starts in the ducts. Love is funding research she hopes might sort out these mysteries.

After we met the first time, Love sent me a medical journal article published in 1977 about women who lived in fishing villages in China and traditionally breastfed on only one side. The authors had studied rates of breast cancer among these women and found that of women fifty-five and older, 79 percent of breast cancer cases occurred in the breast that had not been used for nursing. The study suggested a curious connection between breastfeeding and breast cancer as yet unexplained in the medical literature. The research on the basic biology of breasts is so thin that Love is going back forty years looking for insights. "I think a part of it is sexism," Love said. "It's not malicious, but I think it just doesn't occur to men."

With collaborators, Love's foundation is also testing self-reading portable ultrasound machines that could be used to evaluate palpable breast lumps in poor countries where modern technology and highly trained radiologists are in short supply. The machines are being tested in Guadalajara, Mexico, and are programmed through artificial intelligence to ignore growths that are clearly benign, like cysts. Love told me that, due to limited resources, the wait time for a diagnostic mammogram in some Mexican communities can be as long as nine months. She hopes that by eliminating some of the demand for this more expensive testing, resources can be redirected to women whose lumps are truly suspicious.

When I met Love, she had the freckled face and short curly dark hair I had seen in her old UCLA headshot, but she was thinner. In 2012, Love was diagnosed with acute myeloid leukemia, a particularly deadly cancer in older people. She survived but says the treatment she endured left her with no sense of taste or smell and numb toes. The chemotherapy she received also affected her brain function, leaving her with what she described as a "fog" that sometimes makes it hard for her to remember words and names. This happened a few times during our interviews, prompting Love to scribble reminder notes on a Post-it. (When she later remembered what she wanted to say, she sent me an e-mail.) When Love needs to concentrate on something complicated, like a grant application, she works in the morning, when she is least tired, and drinks a lot of coffee, which helps her stay more focused and clearheaded.

Love said her cancer experience made her more attuned to the "collateral damage" of breast cancer treatment. One of her foundation studies examined how such damage affects the lives of metastatic breast cancer patients. "One big surprise to me—and I thought I was so tuned in—was that people who have the worst collateral damage are those with bone mets," Love told me. Bone metastases, in which breast cancer cells invade the marrow and form tumors in the spine, ribs and other areas, can be enormously painful. But unlike metastatic spread to the brain or liver, bone mets won't kill a woman, so doctors don't always take them as seriously as they should, Love said. Some breast cancer patients suffer significant short- or long-term side effects from chemotherapy and other drugs. These can range from "brain fog" to numbness in fingers or toes to permanent hair loss or severe bone pain. No one knows precisely why some patients suffer these side effects and others do not, but Love wonders if genetic factors might be to blame. She's hoping to spur research in this area.

Love cofounded the National Breast Cancer Coalition, one of the influential advocacy groups started in the 1990s when Komen was growing rapidly. I asked her if she thought Komen might be on an unstoppable trajectory toward irrelevance. "The average person in the street—if you asked them about breast cancer, they still think pink and they think Komen," Love said. Still, she said Komen and other large breast cancer advocacy groups and charities risk alienating younger generations of women if they do not update their messages and research priorities. "A lot of the advocacy is still stuck in the 1990s," she said. The generation of women that blasted Komen for its Planned Parenthood decision is growing older and more at risk for breast cancer every year. When the time comes, will they turn to Komen as women before them did?

On a misty March morning in 2018, I decided I needed to visit the wellspring from which so much pink and consternation had bubbled up and over. I found the starting line of the 2018 Komen Los Angeles County Race for the Cure in the parking lot of Dodger Stadium. From there, the 5K route would loop through a surrounding city park with views of downtown and the Hollywood sign. The race itself started at nine a.m. I arrived at seven thirty. The parking lot was bustling.

My first pink sighting was, of course, a boa. It was wrapped around the neck of a middle-aged woman wearing a fuchsia tutu and pink-and-black-striped tights. Continuing through the parking lot, I saw pink knee socks and pink wigs. Pink capes and pink crowns. Pink hula skirts and pink sunglasses. And then there was the "merch," much of it concentrated in the Komen Store, a large tent with clear plastic walls. At the store, race participants could buy visors, ball caps, mugs, flashlights, flip-flops, shoelaces, pens, coasters, notebooks, Christmas ornaments, beer koozies and, of course, T-shirts. Even though most of the Komen Store merch was pink or partly pink, items were adorned with Komen's new "More than pink" slogan.

I did see more than pink. I saw purple, the color of Komen T-shirts printed specially for stage IV patients, now a common presence at Komen events. ASK ME ABOUT METASTATIC BREAST CANCER, the T-shirts read, along with the word THRIVER. In an area reserved for race participants with metastatic disease, fact sheets about stage IV breast cancer were stacked in a pile, and a poster about breast cancer that spreads to the brain was blown up. Above the race's starting line was a banner that said BE MORE THAN PINK.

Before the race began, I strolled through the rows of sponsor tents set up in the parking lot, stopping to pet a German shepherd dressed in a Komen T-shirt. There was a booth sponsored by Dippin' Dots and two hosted by local supermarket chains that offered free face painting. Some fifty companies, including a nail salon, a makeup company and ShoeDazzle, had made in-kind donations to the event. Two mammogram buses advertising mobile breast cancer screening were parked across from large tables where hundreds of bananas and apples were available for racers who needed a snack.

The criticism Komen has faced for being celebratory made sense to me when I attended the Los Angeles race. The atmosphere at the event was upbeat, with music playing and women dancing, walking arm in arm or posing in an on-site photo booth where there were pink boas and crowns to put on if they had not brought their own. I met a woman wearing a one-of-a-kind blanket around her shoulders. The blanket was printed with photographs of previous races the woman had attended, seventeen in all.

Shortly before the race started, it began to rain. Some racers and walkers wore plastic ponchos, while others huddled together in the cool morning air. Preselected speakers climbed on a small stage by the starting line to talk about their experiences with breast cancer. The last was a patient with metastatic disease who had been living with breast cancer for ten years. After she told her story, she finished by saying, "We need to be a whole hell of a lot more than pink!" A buzzer rang out and the crowd moved forward together.

5

lady parts

In 1876, George Beatson was a medical student at the University of Edinburgh. One day, he got an unusual request. Could he spare the time to care for a wealthy Scotsman suffering from mental illness? The man lived on an estate and needed a temporary live-in caretaker. Beatson was nearly done with his studies and so he agreed. The job would leave him with enough free time to conduct research for a thesis paper he needed to write. Once on the estate, Beatson heard that lambs were being weaned on a nearby farm. That would make as good a thesis subject as any, so Beatson launched a research project on lactation. What he learned would eventually reshape the world's understanding of breast cancer.

Beatson noticed a striking similarity between the sheep he was studying and what he had read and observed about breast cancer as a medical student. During and after pregnancy, Beatson noted, cells proliferated inside the sheep's milk ducts and produced milk. In a cancerous human breast, cells proliferated too, breaking through the duct walls and invading the surrounding tissue. "In short," Beatson wrote in a paper he published on the subject years later, "lactation is at one point perilously near becoming a cancerous process if it is at all arrested." Beatson had stumbled upon a curious connection. It appeared that whatever physiologic force drove milk production also drove breast cancer. But what was the force?

Beatson gleaned a clue when he learned how dairy farmers kept their cows' milk supply high, even after their calves had been weaned. Scottish farmers, knowing that animals do not produce much milk if they are menstruating, made sure their cows became pregnant again shortly after giving birth. In Australia, Beatson learned, some farmers used a different technique, surgically removing the cows' ovaries altogether, stopping menstruation completely and prompting the animals to produce milk practically indefinitely. Both methods, wrote Beatson, "pointed to one organ holding the control over the secretion of another and separate organ." Lactation, a process that starts with cells proliferating in the breast, was controlled by the ovaries. Could these female sex organs also fuel breast cancer? Could removing the ovaries of a breast cancer patient halt the growth of her disease?

It was an intriguing and provocative idea, but after Beatson finished medical school and set up a practice in Glasgow, he abandoned his study of the link between ovarian function and breast cancer. Other doctors were studying cancer causes and treatments, and Beatson figured he could put his skills to better use in other ways. But then years went by with no progress in the treatment of breast cancer. Many doctors believed the disease was caused by some kind of parasite, but they hadn't been able to prove it. Other theories also dead-ended. Surgical removal of small, local tumors remained the only treatment available to women. In cases where cancer had spread, doctors had nothing to offer. This frustrated Beatson so much that, nearly two decades after his farm-lactation research project, he returned to the subject. All he needed was a patient willing to help him prove or disprove his theory that breast cancer was driven by ovarian function.

The woman he found was thirty-three years old, married and the mother of a toddler and a baby. She had noticed a small hard lump in her left breast while nursing her first child, but it didn't hurt, so she ignored it. After she gave birth to her second child, though, the lump began to grow rapidly. By the time she arrived at the Royal Glasgow Cancer Hospital for treatment, the tumor had grown to measure five by three and a half inches. It had broken through the skin and formed a visible wound. The woman underwent a mastectomy and a surgeon removed the lymph nodes around her breast, but within a few months the cancer recurred, forming hard spots that jutted out from the skin along and around her surgery scar.

A colleague asked Beatson to examine the woman. When he met the patient, he pitched her his idea. Let me remove your ovaries and fallopian tubes, he told the woman. I don't know that it will help, but I have a hunch. "She readily consented that I should do anything that held out any prospect of cure, as she knew and felt her case was hopeless," Beatson later wrote. The effect of the surgery on the woman's breast cancer was remarkable. Five weeks after the operation, her recurrent tumor had shrunk. Within four months, the skin around the woman's mastectomy scar was flat and smooth. Eight months later, her cancer seemed to have disappeared entirely. Beatson published a report of the case in the *Lancet* in 1896, urging his fellow cancer doctors to continue the study on the theory that, in cases of advanced breast cancer, surgical removal of the ovaries might slow or even cure the disease. One London physician reported that of forty-six breast cancer patients who underwent ovary removal between 1896 and 1900, more than one-third benefited from the procedure. This was encouraging, but also vexing. In some women, there seemed to be a clear connection between breast cancer and ovarian function. In others, the operation had no effect.

Beatson's work had led to two discoveries. He proved that breast cancer and ovarian function were related, pioneering a treatment that could help some women who had previously been sent home to die. But, just as important, the doctors who followed Beatson's lead produced evidence showing that not all breast cancer was the same.

The reason why would not be understood until the late 1950s, when Elwood Jensen, a chemist at the University of Chicago, happened upon an experiment done by one of his colleagues. The scientist had given prepubescent lab rats minute amounts of estradiol, the primary female sex hormone. The hormone made the rats' reproductive organs grow to six times their normal size. Jensen was fascinated. Other scientists, studying how bodily tissue changed in response to female hormones, had concluded that estradiol interacted with enzymes and transformed in a way that affected the reproductive organs. In a 2010 interview, Jensen (who died in 2012) said he reviewed this conclusion and had another thought. "Maybe we should attack this from a different angle."

Rather than study how tissues were changed by estradiol,

Jensen decided to study the hormone itself. He and a postdoctoral fellow named Herb Jacobson engineered a way to bind estradiol to a radioactive tracer, and they launched a morbid experiment. They injected the substance into young, nonreproducing rats, killed the animals at various intervals over the next day, and measured the rats' radioactive estradiol levels. At first, the substance in the rats' bloodstreams ran throughout their entire bodies. But as the hours passed, all but the estradiol in the reproductive organs was eliminated. Something was making estradiol stay put—a receptor of some kind, a sort of baseball glove inside each reproductive cell that caught the hormone.

Thanks to the work of Beatson and others, Jensen already knew that some breast cancers were fueled by ovarian function. Jensen surmised that these breast cancers must contain receptors similar to those found in reproductive organs. Like beacons, hormone-related breast cancers were attracting estrogen and using it as a kind of fuel. Jensen and some colleagues developed a screen that could determine whether a breast cancer tumor contained the estradiol receptor. Some did, but others did not, solving the mystery of why not every breast cancer patient's disease responded to surgical removal of her ovaries. Only breast cancers that contained the receptor were fueled by estrogen. For these women, removing their ovaries cut the main fuel source feeding their breast cancer. For women whose disease did not have the receptor, ovaries had nothing to do with it.

The gorgeous Los Angeles weather made Eleonora Ford want to run. She ran along the streets of Santa Monica and a paved beach path in Venice. Ford loved the rhythmic sound of her sneakers hitting the pavement. "It was just the switching off and going into the zone," she said. "It was a relaxing thing. It became something I had to do, like eating or sleeping."

Ford had earned a doctorate in chemistry from King's College in London and moved to California in 2001 to take a postdoctoral position in pharmacology at the University of California, Los Angeles. She loved running in the Southern California sunshine so much that she decided to compete in the Los Angeles marathon. She trained with a local running club and finished the race in a respectable five hours, but she knew she could do better. In the years that followed, she ran five marathons in all, crossing the finish line in a San Diego race in under four hours. She even tried trail marathons, running long distances through the woods and up mountains. But marathon running is hard on a body, so eventually Ford switched to triathlons, which combine shorter runs with swimming and bicycling. She trained hard for these races too. On a typical day, Ford would wake up at three thirty a.m., swim two miles at a local college pool and run or cycle after work. On many weekends, she would ride her bike a hundred miles.

Ford's husband, Jo, also loved to sweat and compete. A computer engineer, he had been a semiprofessional soccer player back in England. Ford and her husband had started dating as teenagers. As thirtysomethings, they thought about starting a family but kept putting it off. Ford had taken a job at a biopharmaceutical company just north of Los Angeles. With her work schedule and their busy lives, there always seemed to be a reason to wait on babies. But then Ford's father was diagnosed with terminal lung cancer, and the couple decided they shouldn't delay any longer. Their son was born in May 2012.

From the start, he was a mommy's boy. He followed Ford everywhere and constantly begged her to pick him up for a cuddle. Ford had worried whether she would be a good mother, but she loved everything about it. When her son was two, Ford and her husband decided he should have a baby brother or sister. Ford had undergone intrauterine insemination to get pregnant for the first time. Now forty, she made an appointment with her obstetrician-gynecologist to discuss another round of fertility treatments. Noticing her age, the doctor recommended Ford get a baseline mammogram first. The procedure is uncomfortable for most women, but for Ford, it was excruciating. "It just wouldn't squish," Ford told me.

When the mammogram results came back a few days later, Ford learned why. Her left breast contained an eleven-centimeter mass. "My whole breast was a tumor," Ford told me. A biopsy confirmed the mass was malignant and that the cancer had spread to Ford's lymph nodes. She would need chemotherapy, a mastectomy and radiation therapy. There would be no more marathons or triathlons or hundred-mile cycling weekends. Even worse, there would be no second baby.

George Beatson may have been the first to recognize that sexual reproduction and breast cancer were linked, but there were hints more than a hundred and fifty years earlier that the two processes might be connected. In 1713, an Italian physician published a treatise on the health of workers in various occupations, noting that nuns had an unusually high rate of breast cancer. No one knew why the sisters were far likelier to die of breast cancer than women in the general population, but in the time since, scientists have learned that the answer lies in their celibacy. A groundbreaking 1926 paper about one thousand women in Glasgow and London found that women had a higher chance of developing breast cancer if they had children later in life, had few children or breastfed their babies only briefly or not at all. Estrogen levels fall inside a woman's body when she is nursing, shortening the period during which a rogue breast cell with an estrogen receptor might be able to multiply.

A 1969 study of more than thirty-one thousand nuns who died of cancer between 1900 and 1954 found that they had higher rates of not just breast cancer but also ovarian and uterine cancer. Having babies helps protect women against cancer. (Not having babies doesn't mean you'll get cancer, however. Janet Lane-Claypon, the British scientist who published the 1926 paper, had no children and lived to be ninety.) After the Industrial Revolution, as reproductive patterns in the West changed, with women having fewer children and giving birth later in life, the incidence of breast cancer increased. Birth control, which came along in the 1960s, exacerbated the problem, but ironically it also provided a solution.

Being Craig Jordan's mother was a dangerous job. There was the time he threw a match at some curtains and set them ablaze. And the time he used an air rifle to shoot flowers in the back garden of his family's home in northwest England. Once, he almost killed himself with chlorine gas after a chemistry experiment in his bedroom went bad, although it was hard for Jordan's mother to be angry. She was the one who had allowed Jordan to convert his bedroom into a scientific laboratory. Something was bound to go awry eventually.

Jordan was not a top student in school, but he had been obsessed with chemistry ever since he fished some old science books out of a neighbor's garage. He pored over the texts, especially the chapters about atoms, molecules and the periodic table, even reading them before bed. His stepfather bought him a chemistry set for his twelfth birthday, which was what led to the near-fatal bedroom experiment. Jordan desperately wanted to pursue a job related to chemistry, but at fifteen he hit a roadblock. In postwar England, students were required to pass five standardized subject-based tests to advance to more rigorous academic study. Those who did not pass the tests were funneled into working-class careers. Jordan passed three of the tests (he aced the chemistry exam) but failed the rest, leaving him two short. In danger of becoming a shopkeeper, Jordan applied to be a lab technician at Imperial Chemical Industries (ICI), a company just a few miles from his house.

ICI agreed to hire him on the condition that he pass the final two exams required to advance his studies. After some cajoling from Jordan's mother, the school headmaster agreed to give her son another chance. He passed the remaining two tests, worked part-time at ICI and proved so adept at science that he taught other students university-level biochemistry and was permitted to work in the school lab alone. "I was just ecstatic," Jordan told me. "I'd be in a laboratory on my own doing stuff. They'd be at lunch!" After he finished school, Jordan enrolled at the University of Leeds. A few years after that, he got a scholarship to earn a doctoral degree in pharmacology.

But first, Jordan wanted to join the military. His father, an American, had fought in the Battle of the Bulge during World War II and his mother's father had been a soldier in both world wars. Jordan signed up with an army program for college graduates—the equivalent of the Reserve Officer Training Corps in the United States—and was sent to America to work on a NATO biological and chemical warfare program. Once he returned to England, he dived into his PhD studies, focusing on drugs that affected reproduction. A few years later, Jordan returned to America to conduct research at the Worcester Foundation for Experimental Biology in Massachusetts. He planned to work under a former ICI scientist there, but the man left for another job just as Jordan arrived. Jordan called his PhD adviser, an ICI scientist named Arthur Walpole, and told him he was in America and needed a research project. What should he do?

In addition to manufacturing plastics and bulk chemicals, ICI was in the birth-control business—or rather it *wanted* to be. The first oral contraceptive pill had been approved for use in 1960, and ICI's pharmaceutical division had developed a new compound that it hoped might be a blockbuster morning-after pill. ICI scientists had shown that its new drug prevented pregnancy in postcoital rats, and the company thought it might function the same way for women. It turns out that rats and humans work differently when it comes to reproduction. The drug, known as ICI 46,474, didn't prevent pregnancy; in fact, it increased fertility in women. The drug was deemed a failure, but Walpole had led the research team that developed the compound and he believed it might have a different application.

After Beatson discovered that removing the ovaries of breast cancer patients could sometimes curb their disease, such surgery (or radiation treatment to the ovaries) became standard care for premenopausal women with advanced disease. Perhaps ICI 46,474 could offer an alternative. Although the drug behaved like estrogen in a woman's ovaries, increasing fertility, it had the opposite effect on breast cells, where it acted like an anti-estrogen. Perhaps the pill could serve the same purpose as surgical ovary removal and extend the lives of some women with advanced breast cancer. The problem was that ICI wanted a blockbuster, and developing a drug for women who were likely on the precipice of death held little appeal for the company. Plus, newly developed chemotherapy regimens were thought to be the future of cancer therapy. But Walpole was insistent. He threatened to retire early unless ICI agreed to continue studying the failed morning-after pill for use as a breast cancer treatment. Craig Jordan, who had met Elwood Jensen at Worcester, was eager to work on a breast cancer project. He could lead the research.

ICI agreed, awarding Jordan a grant that would support ICI 46,474 research for two years. In the lab, Jordan discovered that the compound worked like a key broken off inside a lock, filling up the estrogen receptors in breast cancer cells so that hormones could not get in and fuel the disease. By the time Jordan left Massachusetts in 1974, the compound was being tested as a breast cancer drug in preliminary U.S. clinical trials. In England, ICI gave Jordan more funding to keep up the work. Rats were delivered to his lab at the University of Leeds every week. For three years, Jordan studied the effects of ICI 46,474 on breast cancer cells grown inside the animals. He discovered that ICI executives had been wrong—ICI 46,474 *was* a blockbuster. In fact, women needed even more of the drug than the scientists running early breast cancer trials of the compound initially thought. The ongoing American trials of ICI 46,474 had administered the drug to metastatic breast cancer patients for only a year. Through experiments in his lab, Jordan proved that the drug's anticancer effects were dramatically better if the drug was given for at least *five* years—and not just to women who were dying of breast cancer. It could help early-stage patients as well. Now, for the first time, there was a simple pill women could take, a treatment targeted to their particular *type* of breast cancer. When the drug was taken daily, Jordan and others showed that it dramatically reduced the chances that an estrogen-receptor-positive breast cancer would return months or years later.

Today, ICI 46,474, known as tamoxifen, may be the most commonly prescribed cancer drug in the world. Effective for pre- and postmenopausal women, a compound that was developed as a morning-after pill and nearly discarded has been used to treat millions of women. The drug is sometimes also prescribed as preventive medicine to women without breast cancer who have a strong family history and are at high risk. In 1985, ICI's pharmaceutical division helped start National Breast Cancer Awareness Month in its largest market, the United States. In 1993, the division was spun off into a new company, Zeneca, which in 1999 merged with Astra to form AstraZeneca.

Tamoxifen was revolutionary, not just because of its effectiveness but because it launched a whole new way of treating breast cancer. The work of Beatson, Jensen and Jordan showed that there were different types of breast cancer, each with its own risks and causes. Tailoring breast cancer treatments to the biology of tumors has driven scientific research for the disease ever since.

Tamoxifen works by blocking breast cancer cells' ability to feed on estrogen circulating inside a woman's body. But getting rid of all the estrogen at its sources can be even more effective. More than a century after George Beatson discovered that surgically removing a thirty-three-year-old breast cancer patient's ovaries could halt the growth of her disease, premenopausal women diagnosed with estrogen-receptor-positive breast cancer still sometimes undergo the procedure. Other young breast cancer patients take ovarian-suppression drugs, which shut down a female's main estrogen factory. But even without functioning ovaries, a woman's body is not estrogen-free. An enzyme called aromatase, found in body fat, helps turn androgen, produced by the adrenal glands, into small amounts of estrogen. This secondary source of estrogen becomes especially important after menopause, but it is also able to fuel breast cancer. A relatively new class of drugs shuts down this process as well. Called aromatase inhibitors, the drugs stop the enzyme's function. In head-to-head comparisons with tamoxifen, aromatase inhibitors are moderately more effective at preventing recurrences of estrogen-receptor-positive breast cancer. In a way, they neuter women, removing from their bodies the very chemical that makes them female.

An estrogen-free body is far less likely to develop breast cancer, but it can also be a difficult body to exist in. Aromatase inhibitors can cause severe joint pain in some women and dangerously reduce their bone density, making fractures more likely. As many as one-third of women prescribed aromatase inhibitors stop taking the daily pills earlier than their doctors recommend. Most just can't endure the side effects. Many of the women interviewed for this book who take aromatase inhibitors say they were unprepared for the havoc the drugs would wreak on their lives. Oncologists may be reluctant to dwell on all the possible downsides of the drugs for fear that women might refuse to take them. Not all patients on aromatase inhibitors experience severe side effects, but for those who do, the drugs can redraw the boundaries of daily existence.

Eleonora Ford's eleven-centimeter tumor was estrogen-positive. In addition to other cancer treatments, Ford wanted to have her ovaries removed. Her oncologist urged her to wait, but Ford's mind was made up. She wanted the most aggressive treatment available. After the surgery, Ford's oncologist prescribed the aromatase inhibitor exemestane, which she would take for ten years. When I met Ford one afternoon in her fourth-floor condo on the west side of Los Angeles, she still looked like the athlete she once was. She was wearing a tank top that showed off her sinewy arm muscles and workout stretch pants that revealed her sculpted calves. But these were vestiges of her nearly two decades as an endurance athlete. They bore little connection to her current physical health.

"From being a triathlete to today, it's night and day," Ford told me. "It's really like I've got an eighty-year-old body that's very creaky." Ford had to give up cycling. The repetitive motion of pedaling left her in too much pain. She used to love hiking, but Ford told me the activity causes severe aches in one of her hips. She tried yoga, hoping a gentler form of exercise might be easier on her body, but even that was painful. Kneeling down to play with her toddler son hurt too. "It was excruciating on my knees and ankles," Ford said. "It's very frustrating. I'm not in a negative place about it, but it's just kind of the new baseline, right? A new reality." Like many women on exemestane, Ford also has a new sexual baseline. The lack of estrogen in her body has left her vagina feeling eighty years old too.

"I've been to my gynecologist to ask, 'Is this it? Is this going to get any better? Because this is obscene,'" Ford said. The lack of natural lubrication and the thinning of the skin and tissue in her vagina has made even urinating and wiping painful. "It's sensitive down there, let alone any kind of penetration," Ford said. "It's a horrible thing. I'm lying in bed, and I'm scared because it's going to hurt." For decades, the effect of breast cancer treatment on patients' femininity focused on the loss of breasts, but Ford's experience illustrates how other parts of the female existence can disappear too.

In one of Ford's breast cancer support Facebook groups, women on aromatase inhibitors exchange tips about how to make sex less painful and more feasible. "Time just goes on, and then you lose that connection. And I think for a couple, without that connection, then what are you? Just friends?" Ford's husband has been supportive and understanding about the sexual side effects that came in the wake of her taking the aromatase inhibitor. "But I know that it's hard," she said.

After her primary cancer treatment ended and Ford returned to work, she told me that she was so crippled with anxiety—about her cancer coming back, about the changing nature of her marriage—that her memory started to fail her. "I'd sit in a meeting, and because I was so anxious, nothing would go in my brain," Ford said. "People would talk, and all I could hear was that 'Blah, blah, blah.'" Ford was so worried she would forget something important that she began taking copious notes at work that she would review over and over once she got home. "I would go home and spend hours," she said. Ford remembered chopping vegetables for dinner one night when a kitchen timer went off and sent her into an anxiety spiral. "I wanted to ball up on the floor. I couldn't deal with one more thing," Ford said.

Ford began to have disturbing thoughts. She started to wonder if she had forever lost the ability to be the high-functioning person she had been before cancer. She even wondered if her husband and son might be better off without her. "I just wanted to be gone," she told me. "I've typically been a glass-half-full kind of person. And as someone who's a scientist, I think I'm a very practical, logical person." Ford knew intellectually that her family would be devastated to lose her, but she couldn't shake the thought. Her oncologist had suggested Ford try antidepressant drugs. She resisted, but eventually the dark thoughts overwhelmed Ford and she gave in, filling a prescription for the drug Lexapro. Before long, she began to feel better, more like the mother and wife she had been before cancer.

"Now, with that fog lifted, it's like, 'What was I thinking?' They need me. They'd be terrible without me," Ford said. "I make them smile, and we love each other." Studies show that exercise can help reduce the chances breast cancer will recur. Although she cannot run marathons or take hundred-mile bike trips, Ford has started attending Bar Method classes, a low-impact workout she can modify to suit her body. She takes long walks and makes sure to get a lot of sleep, about nine hours every night. Ford has also been reading up on a relatively new vaginal laser treatment that may reduce pain during sex in postmenopausal women. (Because Ford had her ovaries removed, she is postmenopausal, despite her young age.) Some women who experience severe side effects on aromatase inhibitors switch to tamoxifen, which does not cause bone-density loss or joint pain and can have less severe sexual side effects. Ford won't hear of it. Studies have shown that aromatase inhibitors are slightly more effective than tamoxifen and she's still committed to aggressive treatment, knowing it is her best shot at remaining cancer-free.

The first time I met Ford, she told me she had recently experienced another side effect associated with antiestrogen therapy—tooth decay. "I found out that another tooth—a second one—is dissolving from the root," she said. When we talked a few months later, Ford said the tooth eventually cracked and had to be pulled. She was gearing up for a procedure in which an oral surgeon could insert a pin on which an artificial tooth would eventually be installed. She told me her dental problems had to be addressed before she could begin taking a new type of drug that studies have shown may modestly reduce her chances of getting bone metastases. The drug could also increase her bone density, which has decreased over the years as a consequence of her antiestrogen therapy.

Despite her ordeal, Ford is still a glass-half-full kind of person. In our conversations, she wore a near-constant smile that was infectious. Her singsong British voice frequently broke into laughter, as she told me about the joys of being a mother to a now-school-age little boy. Even though it can be painful, she scoops him up into her arms some days and often chases him around their apartment. Ford said she has no regrets about her treatments and is enormously grateful to be alive, even if life has a new definition.

The side effects of antiestrogen therapy can mimic the symptoms of natural menopause, which range from vaginal dryness and hot flashes to moodiness and sleep disruption. For decades, women experiencing these symptoms as a result of natural menopause were treated with supplemental hormones. Worried the drugs might be causing unforeseen side effects, researchers launched a federally funded national study in 1993 to examine whether the benefits of hormone-replacement therapy outweighed its harms. The study was stopped early, in 2002, after researchers found that a common postmenopausal hormone-combination therapy increased the risk of breast cancer. In the years since, researchers have found that different types of hormone therapy used for short periods may be safe.

Doctors have also worried that hormonal birth control might increase breast cancer risk. The best study of this topic, a Danish paper published in 2017, followed nearly two million women for a decade. It found that hormonal birth control modestly increases the risk of breast cancer and that the increase in risk is greater for women who take the drugs for many years.

Men get breast cancer too, some twenty-five hundred annually. Male breast cancer kills about five hundred men per year in America, more than die of testicular cancer. The disease develops inside the small amount of breast tissue contained in every man's chest. The disease is rare, but awareness is increasing. The Male Breast Cancer Coalition (MBCC), founded in 2014, held its first conference in 2017. A cadre of attendees, many with scars across their chests, posed shirtless for a picture on the MBCC web page promoting the event. There's a blue-and-pink ribbon to signify support for men with breast cancer, a Twitter hashtag—#menhavebreaststoo—and even a Recovery Shirt for sale on the MBCC website; it has internal drain pockets and front Velcro fasteners that are easy to open and close, which is useful if you're a man who's just had a mastectomy.

But 99 percent of breast cancer cases are inextricably tied to motherhood or, more precisely, the potential for motherhood. For decades, most young women who survived breast cancer were no longer able to have children because their ovaries had been surgically removed or because their ovaries were intentionally rendered useless by radiation or hormonal therapy. Chemotherapy can also severely damage the ovaries. Those young women with estrogen-receptor-positive cancer who were still able to conceive after treatment were warned by their doctors to avoid pregnancy. Pregnancy reduces estrogen levels in the body, but there was concern that the dramatic hormonal fluctuations of gestation could somehow trigger recurrences. Relatively recent research has proved those worries to be unfounded. Women who get pregnant after breast cancer have no more risk of recurrence than those who don't conceive after treatment. Of course, getting pregnant after breast cancer is not easy, even for a woman with disease that is not estrogen-receptor-positive. I fell in this group.

And so, four days after I was diagnosed with breast cancer, my husband, Collin, and I found ourselves in the office of Dr. Richard Marrs, a reproductive endocrinologist. We had planned to start trying for another baby at the start of 2015. When I was diagnosed with breast cancer in December 2014, the surgeon who delivered the news urged us to talk to a fertility specialist before we fully plunged into Cancerland.

I barely listened when the surgeon who diagnosed me brought up the subject of future children. Saving my life seemed the only priority that mattered. "Right now, all you can think about is cancer, which is fine," the doctor said. "But one more month to do what is important to you, get your eggs extracted, get the embryos and store them — I don't think you're going to regret it. I think you would regret it if you didn't do it." So there we were, sitting in Marrs's large office above Wilshire Boulevard. (Among other reasons, Eleonora Ford did not want to delay the start of her chemotherapy to have her eggs extracted.)

Marrs had made history in 1986 as the first American doctor to successfully implant a previously frozen embryo. We shook hands and he settled into his chair. "You have a lot on your plate," he said. A doctor who helped create life understood what it meant to fear losing it. I described my diagnosis and concerns about delaying my cancer treatment in order to create and preserve embryos. Marrs said we could immediately start coaxing my body to release an abundance of eggs during my next ovulation cycle, which was just ten days away.

Marrs examined me and sent me home with a paper bag of injectable hormones and a collection of fresh needles. Collin scheduled an appointment to provide a sperm sample. In the week that followed, I pinched the fat of my belly and plunged the tiny needles into my flesh once or twice a day. While we waited for life to change irreversibly, we tried to maintain some normalcy, which meant I wasn't always home when it was time for an injection. Once, we were at a fancy holiday party and I had to retreat to the car to administer the drugs. I pulled my dress up to my midsection, my vials and needles spread out on the dashboard. As a stranger walked by the car, I realized that I looked like a dope fiend. Another time, I was shopping with my mother for a holiday-dinner tablecloth and had to shoot up in the parking lot.

By the time I arrived at Dr. Marrs's office early in the morning the day after Christmas, I was bloated from the hormones and worried we were wasting money Collin might need once I was dead or too sick to work. The idea of having another baby, though, had generated another feeling—hope. Hope that there would be life after cancer, that there would be an "after cancer." We made twelve embryos in all and sent them to a cryobank that charged us eight hundred dollars a year in storage fees. Altogether, my week and a half as a fertility patient cost nearly ten thousand dollars.

6

not your mother's chemotherapy

Cancer has a way of bringing those around you into sharp focus. Watching your husband toil over the stove one evening, still wearing a button-down shirt and an ID badge from work, you might glimpse a level of capability that brings you immense relief but also heartbreak. You might suddenly realize he can work long hours at a high-pressure job and still have the energy to make dinner, plan weekend adventures and read bedtime stories to your kid until she falls asleep. He could be a great single dad, you think. He could do this on his own.

Along the way, you might also realize that some of your closest friends care less about you than you'd thought, like the one who calls once after your diagnosis and never again. Or the one who says she's too busy to visit. Or the one who breezily changes the subject every time cancer drifts into a conversation. Does this person understand you could die? you might wonder.

Other people might turn out to be better friends than you'd thought, like the one you barely know who sews you a traditional

Ukrainian doll she says is imbued with protective powers. Or the new friend who puts your chemo dates on her calendar and texts the morning of each one. The husband of a friend who cooks a celebratory fish dinner the night after a chemo session. The neighbor who finds out about your cancer, gives you her phone number and tells you to call her at work if you ever need anything. Her boss has been apprised of the situation, so it'll be fine, she says. Or the other neighbor who knows you love to grow vegetables and builds you a raised garden bed made of redwood.

If you are lucky, you will receive all sorts of weird gifts when you are a cancer patient: Artisanal canned tuna and a bag of bendy straws. A pair of bright blue sneakers. Wind chimes. Lotion, so much lotion. A Larch Wood cutting board that feels like velvet.

Then there are those who don't surprise you with their generosity, although you are very grateful for it. The best friend from high school who travels across the country for two of your chemo infusions to take you shopping and watch reality TV with you. The little brother who is a public-school teacher and maxes out all his credit cards so he can fly from New York to California four times in five months. The aunts, trained nurses, who travel thousands of miles to take care of you. One doesn't like to fly, but she boards two planes for the trip anyway.

During the five months I underwent chemotherapy, as my friends and family drifted in and out of my vision, time itself became a series of boundaries and borders. My infusions took place every third Thursday. I quickly learned to plan my life around the number of days since my previous infusion, the number of days till my next one and the number of days until

I would be done with the whole thing. The calendar provided predictability. Here's how it went:

> Wednesday: Take steroid pills to ward off an allergic reaction to tomorrow's chemotherapy. Take antianxiety pills to ward off the jitteriness and insomnia caused by the steroid pills.
>
> Thursday: Get chemo.
>
> Friday: Take steroid pills to prevent an after-chemo allergic reaction. Take antianxiety pills to ward off the jitteriness and insomnia caused by the steroid pills. Return to the clinic for a shot of a drug that will boost my immune system. Begin to feel sick. Take anti-nausea drugs. Smoke pot.
>
> Saturday: Feel sicker. Take anti-nausea pills. Smoke pot. Sleep.
>
> Sunday: Feel sick. Take anti-nausea pills. Smoke pot. Sleep.
>
> Monday: Feel somewhat sick. Take anti-nausea pills.
>
> Tuesday: Feel better.
>
> Wednesday: Return to work.

Two weeks would pass and I would start the cycle over again. Our friends, meanwhile, celebrated the forward progress of their lives. They got married, started new jobs in new cities, had babies, bought houses. The milestones were normal for people our age, but I resented them nonetheless. While my friends were living, I was trying not to die.

One night between infusions, I dreamed that I *had* died.

In the dream, I was lying on a table as formaldehyde was pumped into my veins. I was trying to scream that I wasn't ready, but no sound came out. I mercifully woke up, breathing fast as my heart pounded in my chest. It was just a dream, I told myself. You're not dead. Another night, Evie had a fever and crawled into bed with Collin and me. As predicted, chemotherapy had shut down my ovaries and kicked me into chemical menopause, which caused me to have hot flashes around the clock. Evie and I sweated together all night.

"At least once a day, I think about you getting a brain tumor," Collin told me one afternoon, leaning against the kitchen sink. We had promised each other that nothing about cancer was off-limits and I appreciated his honesty, that he let me know what he was thinking. I was thinking the same thing. Although I was receiving a treatment protocol that worked for most women like me, it was hard not to indulge in thoughts about being on the wrong side of the statistics again. We would not know how well the chemotherapy was working until I underwent surgery afterward and the doctors could look inside my breast. My nipple discharge stopped after my first chemotherapy session, but it returned a few months into treatment, clearer and thinner than before. My oncologist surmised that the change was a reassuring sign indicating that my tumors were dying.

Nancy Lankford's love affair with trees started when she was a child. Her family vacationed every summer in Washington State's San Juan Islands, where the rocky, dry coastline is the perfect environment for Pacific madrone trees, a tropical-looking species with peeling red bark. Lankford was awed by their twisted trunks and long branches, which hung over Puget Sound. On weekends, Lankford's family hiked in Mount Rainier National Park, an area dotted with spire-like subalpine firs, and in the Olympic Mountains' Hoh Rain Forest, which was crowded with western hemlocks and western red cedars.

When Lankford was eight, her parents delivered some devastating news—they had to cut down a gnarled old apple tree in the backyard to make room for a house addition. Lankford had climbed the tree nearly every day. She was distraught. To ease her anguish, her parents got her a replacement apple tree for her next birthday. "I still remember that as one of my favorite birthday presents," Lankford said.

In high school, Lankford joined a mountaineering club in Seattle and went on an overnight backpacking trip. It changed her life. "I became a backpacking and mountain-climbing fool," she said. After that first trip, Lankford spent pretty much every weekend in the woods. She loved the stillness of the forest and the puzzle-piece-like bark of ponderosa pines that smelled sweet in the sun. The summer after Lankford graduated from high school, in 1977, she signed up to be a volunteer wilderness ranger with the U.S. Forest Service in Mount Baker-Snoqualmie National Forest near Washington's border with Canada.

By the time Lankford got to college, she knew she wanted to study something that would keep her amid the trees of the Pacific Northwest. She wanted to protect the land where she had hiked and camped all her life. But as Lankford studied the forests, her thinking changed. She came to understand that the swaths of land she fell in love with could provide more than just peace and scenery. "Wood is one of our most sustainable resources," she said. Lankford became a forester in 1982, when women in the job were not common. At parties, when people asked what she did for work, she had a stock answer: "I'm a tree hugger, but I cut 'em down too."

Lankford's specialty in the U.S. Forest Service was silviculture, the study of trees. After a few years working on timber-harvesting projects, reforestation and forest-fire management, Lankford was promoted to oversee a genetics program in the 1.1-million-acre Mount Hood National Forest in Oregon. She selected and grew new trees based on traits that would yield high-quality timber in large quantities and yet maintain the forest's biodiversity. Then one day in the fall of 1990, Lankford's boss said she had been reassigned. Instead of working on the genetics program, Lankford would help steer a federal effort to harvest bark from a tree no respectable timber company would ever want.

The tree was a squat species called *Taxus brevifolia,* otherwise known as the Pacific yew. "It's not the biggest, fullest, tallest tree," Lankford told me. "It's kind of like the underdog." In contrast to the towering Douglas fir trees that were made into building lumber, the Pacific yew was a "trash tree." It grew no higher than fifty feet and lived beneath the forest's canopy in the dark, damp space nearest the ground. There wasn't a lot of room or light down there, and the cramped environs often caused Pacific yew trees to bend themselves into contorted shapes. Sometimes one grew straight enough to be made into a few fence posts or archery bows, but more often the trees ended up in slash piles of waste that were incinerated in controlled burns.

Lankford's bosses weren't after the Pacific yew for timber.

Rather, they wanted its bark, which contained a toxin that scientists on the East Coast had learned how to distill into a potent medicine. The medicine was a new type of chemotherapy, and it appeared to be an effective agent against breast cancer cells. Clinical trials were already under way to test the yew-derived drug's efficacy, but supplies were running short and the doctors delivering the experimental therapy needed more trees to make more medicine—fast.

"It was late fall and this was a 'hurry up and do it now' kind of thing because we had to get out there before the snow," Lankford said. "I was all in. Here was a product the forest could provide. I've taken a lot of walks in the woods for different reasons. Sometimes you feel like you're just counting trees. But this was exciting and important work."

Lankford understood her new assignment might mean the difference between life and death for women with breast cancer. That she would someday be one of those women never occurred to her.

The first chemotherapy drugs were products of war. During World War I and World War II, some American service members were exposed to chemical weapons, including mustard gas. Doctors who performed autopsies on the men who died noticed something odd—the soldiers' bone marrow had been particularly devastated by the gas. Their white blood cells, which are made in the marrow, had been nearly wiped out. The mustard gas acted with such precision that scientists later found it could be used to treat leukemia and lymphoma, both cancers of the white blood cells.

By the 1950s, six chemotherapy agents were approved in

the United States to treat cancer. All were synthesized by scientists in laboratories. For the first time, the drugs allowed doctors to fight cancer systemically, sending treatment all over the body through the bloodstream. Surgery, radiation and chemotherapy constituted a new trifecta that offered a fresh promise of cure for people with some of the most virulent forms of the disease. The new chemotherapy drugs were effective against some cancers, especially leukemia and lymphoma, but for many cancer patients, including those with breast cancer, existing chemotherapy drugs fell short. Hoping to spur development of new agents, in 1955 the National Cancer Institute launched the Cancer Chemotherapy National Service Center. Five years later, the NCI started an ambitious program to look outside the lab, in the natural world, for anticancer compounds. (A juniper bush and a flowering periwinkle already looked like they might be useful in developing medicines to fight cancer.) The NCI hired botanists to collect plants from across the United States and outside the country. Samples were bagged and sent back to labs that tested the specimens to see if they affected cancer cells. Scientists suspected that the natural world contained cancer-fighting compounds, but they had few leads, so the NCI botanists collected samples randomly, hoping to get a hit. Some thirty thousand plant samples were screened between 1960 and 1981. One looked particularly promising. It was collected by a Harvard-trained botanist named Arthur Barclay.

Barclay worked for the U.S. Department of Agriculture, which was collaborating with the NCI on the project to search for anticancer compounds in nature. He traipsed through forests, collecting specimens for the USDA in South Africa, Mexico, and the Western United States, gathering seeds, clipping leaves and scraping bark from trees.

In August 1962, Barclay was on the last leg of a collection trip, walking through Washington State's Gifford Pinchot National Forest with some students, when he spotted a twenty-five-foot Pacific yew growing at about fifteen hundred feet above sea level. The tree had a reddish bark, glossy needles and small, salmon-colored berries. Barclay peeled some bark from the tree, grabbed some of the tree's needles, put his loot in a bag, and sent it to a lab working with the NCI. Scientists there tested the bark and found that it was toxic to living cells. In 1964, the NCI asked Barclay to return to the site of his initial collection and retrieve a second, larger sample. It would take years to determine if the bark toxin would work as an anticancer drug. By 1970, scientists had figured out the bark compound's complex molecular structure. The next year, President Richard Nixon declared in his State of the Union address that the federal government was launching a "War on Cancer." Congress passed and Nixon signed the National Cancer Act, which sent millions more dollars flowing into federal cancer research.

In the late 1970s, a molecular pharmacologist at the Albert Einstein College of Medicine in New York City figured out precisely how the compound, which was named Taxol, worked to destroy cells. (A team of NCI scientists had also determined the chemical's mechanism of action.) Most chemotherapy drugs developed in the lab induced cancer cells' death by damaging DNA or inhibiting specific enzymes essential to cell division. Anticancer drugs derived from plants, however, killed cancer cells by hampering their ability to create internal structures made of tiny filaments called microtubules. Taxol did the opposite—it prompted cells to create microtubules, but in a way that stopped a cell from constricting and changing shape during cell division. The discovery was a major breakthrough. Lab and mouse experiments seemed to show that Taxol was particularly effective against metastatic breast and ovarian cancer, and scientists had finally figured out how to distill Pacific yew bark toxin into a soluble form that could be delivered intravenously. To know for sure, they had to administer the drug to patients. The NCI launched phase 1 clinical trials of Taxol in 1984.

But there were some obstacles to overcome. First, there was the tree itself, which was small in stature; it could take as many as six Pacific yew trees to make enough medicine for a single patient. Stripping a Pacific yew's bark is deadly for the tree, which takes about a hundred years to reach maturity, so quickly producing more bark would be impossible. Plus, Pacific yews were not grown in nurseries or neat rows. They were scattered all over federal lands, hidden among other types of trees. No one knew exactly how many Pacific yews existed in the forests. Finding, cutting and stripping the trees one by one would never yield enough bark to produce all the Taxol that doctors predicted they would need every year for women diagnosed with ovarian and breast cancer. One alternative was to clear-cut the forests, hacking down all trees in a given area and sorting the Pacific yews from the rest. This was unpalatable for environmentalists newly committed to protecting the centuries-old forests of the Pacific Northwest, home to the threatened spotted owl among other species.

But the NCI couldn't give up. By the 1980s, several chemotherapy regimens had been shown to work against breast cancer, but many women still saw their disease return and become resistant to existing treatments. Plus, the chemotherapy cocktails in use were highly toxic and sometimes caused heart damage. The idea that Taxol might work better than these other drugs and have fewer side effects was too appealing to ignore. By the late 1980s, NCI-contracted tree harvesters had managed to collect enough Pacific yew bark to supply the first round of human clinical trials. A small study at Johns Hopkins University showed that one-third of ovarian cancer patients whose tumors had not responded to other types of chemotherapy had major responses to Taxol. At M. D. Anderson Cancer Center, a small trial of women with metastatic breast cancer found that half the patients saw their tumors shrink considerably. A few women saw their masses disappear entirely. Other trials were already under way, but the doctors guiding them quickly ran short on Taxol, which slowed down the research.

In 1989, when Nancy Lankford was reassigned to the Pacific yew harvest, there was no time to lose. "The message was that the women in those trials were women who did not respond to any other chemical treatment," said Lankford. "I thought it was such a unique opportunity—being a bridge between forestry and medicine." Lankford's first job was to count. She helped compile a Pacific yew tree inventory in Mount Hood National Forest, mapping the known trees and measuring their heights and diameters. Locals in Oregon who had permits to cut trees in the national forest—for timber, boughs for Christmas decorations, or other uses—were recruited en masse and hired to cut the trees and strip their bark. Protesters showed up at some of the harvests, charging the federal government with environmental destruction and irresponsible land management. The criticism stung Lankford, who had devoted her life to the Pacific Northwest's forests.

Pharmaceutical giant Bristol-Myers Squibb entered into a cooperative agreement with government laboratories to harvest and synthesize Pacific yew bark from federal lands, removing nearly five million pounds of raw material from the forests over several years. To ensure the process was environmentally responsible and not wasteful, Congress passed the Pacific Yew Tree Act in 1992, which, among other provisions, banned the trees from being burned in slash piles. The harvests Lankford helped coordinate produced enough Taxol to keep the clinical trials in business. The studies proved definitively that the drug worked for breast and ovarian cancer that had not responded to standard chemotherapy. The Food and Drug Administration approved Taxol for these patients in 1992 and for metastatic breast cancer patients in 1994. But the race was on to find an alternative raw material, one that wouldn't draw the ire and protests of environmental groups and, crucially, would be sustainable. Doctors were eager to see if Taxol might work for women with early-stage breast cancer as well. For this use, they would need far larger supplies of the drug. Two months after Taxol gained its first FDA approval, Bristol-Myers Squibb announced the Pacific Northwest yew harvest would wind down. The company had discovered that a faster-growing yew species, found in Europe and Asia, could be used to produce a drug very similar to Taxol. European and American scientists, meanwhile, figured out how to create semisynthetic versions of the drug in the lab.

In the fall of 2011, Lankford was diagnosed with estrogen-receptor-positive breast cancer. "I actually found the lump myself," she told me. She was treated with docetaxel, the drug developed from the bark of the European yew tree. Lankford finished her treatment in mid-2012, a few years before she became a grandmother.

Arthur Barclay's trek through the woods of Washington in 1962 was a long shot that has saved countless lives. The discovery of Taxol also eventually established a new treatment paradigm for breast cancer, demonstrating that less-toxic therapy, which Taxol was when compared with previously existing chemo drugs, could work as well as harsher treatments. But that kind of thinking came later. Right around the time the FDA approved the Taxol chemotherapy treatment for breast cancer, another trend was taking hold. Oncologists thought they had found a way to beat back even the most dire cases of breast cancer without any new drugs at all.

Scientists knew that chemotherapy could kill billions of breast cancer cells, but too often some of those cells managed to escape the onslaught. This was especially true in women whose disease had spread to ten or more lymph nodes before it was discovered. Once the disease progressed that far, there was often little that modern medicine could offer in the way of a cure. Part of the problem was chemotherapy's toxic nature. Indiscriminate and imprecise, the drugs were designed to kill rapidly dividing cells. These included, in addition to cancer cells, the cells on the inside of the digestive tract, those that controlled hair growth and those that made up a patient's spongy bone marrow. An upset stomach and a bald head are survivable inconveniences, but bone marrow is vital to life. It produces red blood cells and platelets as well as the white blood cells that fight infection. Giving more than the standard dose of chemotherapy to a woman might kill her breast cancer once and for all, but it would wipe out her white blood cells beyond the point of recovery, which would also cause her death.

But maybe there was a way to give breast cancer patients higher doses of chemotherapy *and* keep them alive. After all, it had been done in lymphoma and leukemia. To kill off those cancers, doctors had developed a new protocol. They delivered high doses of chemotherapy, bringing lymphoma and leukemia patients to the brink of death, and followed the treatment with another substance—healthy bone marrow. The procedure, in which bone marrow was extracted from the patient before chemotherapy or provided by a donor, allowed doctors to rescue patients after delivering high doses of toxic chemotherapy drugs. In 1979, Gabriel Hortobagyi, a Hungarian-born Colombia-educated oncologist at M. D. Anderson Cancer Center in Houston, decided to try the method out on a breast cancer patient. A young Dutch doctor had arrived at M. D. Anderson a few years earlier and was developing clinical trials for high-dose chemotherapy and bone marrow transplantation in leukemia patients. He knew the protocol and shared it with Hortobagyi.

The procedure seemed to shrink tumors in breast cancer patients, and in the years that followed, Hortobagyi tried it again and again. In Boston, a small group of doctors at the Dana-Farber Cancer Institute had the same idea, beginning high-dose chemotherapy treatment with bone marrow transplantation in the early 1980s. A breast cancer patient there, a truck driver, had gone through standard chemotherapy treatment that failed to kill off her disease. She became the first patient in Boston to undergo a high-dose chemotherapy regimen and bone marrow transplant. She survived the procedure but died within a year. Still, Boston doctors were encouraged. One large tumor in the woman's chest had disappeared in the wake of her chemotherapy treatment, which was ten times the normal dose.

After the Boston doctors tried the procedure on more patients with seemingly promising results, one oncologist on the team, William Peters, decided to leave Dana-Farber for Duke University Medical Center, where he could experiment further with the procedure and study it full-time. Meanwhile, word spread that a new method for treating the worst cases of breast cancer might offer the cure that had so far eluded women and doctors. By the late 1980s, the number of breast cancer patients receiving high-dose chemotherapy followed by bone marrow transplants annually numbered in the hundreds, with tallies increasing every year. Then, in 1993, Peters and his colleagues published the results of a study that upped the ante. In Peters's trial, eighty-five breast cancer patients underwent high-dose chemotherapy followed by bone marrow transplantation between 1987 and 1991, and researchers found that they had a lower chance of relapse than women who'd undergone traditional chemotherapy. Based on an analysis, Peters said the chance of remaining cancer-free at a median follow-up of two and a half years was 72 percent. (At the time, 38 to 52 percent of breast cancer patients whose disease had spread to at least ten nearby lymph nodes saw their disease recur within two and a half years.) In the article, Peters cautioned against overinterpreting the trial results. The follow-up period had been short, and, rather than a randomized trial—in which patients were arbitrarily and equally assigned to standard chemotherapy or the high-dose regimen—the study had merely looked at a high-dose group of patients. Plus, ten women had died of complications caused by the treatment itself.

But the news that doctors had discovered a possible cure for some of the country's most hopeless breast cancer cases had already reached not just oncologists but women themselves. Even breast cancer patients whose disease had spread to fewer than ten lymph nodes, the threshold for Peters's study, were demanding the treatment. Many insurance plans refused to cover the procedure, which could cost more than a hundred thousand dollars per patient, arguing that it was experimental, but patients and their families sued and won case after case in court. In 1993, a jury awarded eighty-nine million dollars to the family of a deceased breast cancer patient whose insurer had refused to cover a high-dose chemotherapy procedure. Sometimes, insurers relented and paid for the procedures, even in the absence of hard evidence that it worked. The negative press they received for not covering the regimens—as well as the threat of legal liability—was not worth the money they would save by refusing coverage. In 1994, the federal government decreed that government insurance plans must cover the procedure. The Blue Cross and Blue Shield Association had grown so weary of losing court cases over the issue that it had already agreed to help pay for randomized trials to see definitively if the procedure worked. Peters persuaded the National Cancer Institute to spend millions to study the issue through trials, but enrollment was sluggish. Why take a chance of being consigned to standard chemotherapy when you could be guaranteed to get the newer, high-dose treatment?

Finally, after five years, four trials from American and European research groups had collected enough patients and data to determine whether high-dose chemotherapy was superior to standard regimens. A researcher in South Africa who had completed an earlier study that showed promising results from the high-dose treatment added a fifth set of data. The cancer community anxiously waited to learn the results of the studies, which would be unveiled at a meeting of the American Society of Clinical Oncology (ASCO) in 1999. Hortobagyi, the M. D. Anderson doctor who had first tried high-dose chemotherapy treatment on a breast cancer patient in 1979, was among those watching closely. He had conducted one of the first clinical trials testing the efficacy of Taxol and knew that controlled, randomized studies were the only way to learn if a cancer treatment that seemed logical in theory actually worked in practice. Hortobagyi had even published a paper the year before the ASCO conference that cast doubt on the efficacy of the high-dose regimens. To qualify for and withstand a high-dose chemotherapy regimen, which killed some people and made the rest deathly ill, breast cancer patients had to be otherwise healthy and strong. Many women who underwent the procedure were young. The median age of women diagnosed with breast cancer in the United States is sixty-two, but in Peters's trial at Duke, only three of the eighty-five women treated had been over fifty. Perhaps healthier patients were being steered toward high-dose regimens, skewing the results, which were compared against historical data collected on patients of all stripes who had undergone standard chemotherapy. Plus, the procedure was dangerous, with up to 10 percent of patients dying of infections or other complications. The randomized trials would provide clarity.

A few months before the big ASCO meeting, Hortobagyi was invited to South Africa to give a talk at a cancer conference on the high-dose chemotherapy experience in the United States. By now a skeptic worried that the treatment had become popular without enough solid evidence, Hortobagyi voiced his concerns. He noticed a man scowling in the front row. The man was a South African oncologist named Werner Bezwoda, whose study of high-dose chemotherapy would be presented at the ASCO meeting.

Finally, the day came for the five studies to be presented at the society's meeting in Atlanta, where some twenty thousand cancer experts from around the world had gathered to hear the latest in oncology research. A large room at the Georgia World Congress Center had been set aside for the presentation, but it wasn't large enough. Two additional conference halls held an overflow crowd that watched the presentations via video feed. Hortobagyi, who had not been involved in the randomized trials, was onstage to provide commentary. In the American and European trials, data showed that breast cancer patients who underwent high-dose chemotherapy fared no better than those who were treated with standard or moderately high chemotherapy. Patients relapsed at about the same rates. But the South African trial had turned out differently. In a study of 154 women, Bezwoda reported that women who had undergone high-dose chemotherapy relapsed at less than half the rate of women in the control group. Bezwoda was the only author of the South African study abstract presented to ASCO, a fact Hortobagyi found odd. "Typically, there are fifteen to twenty authors on these types of papers," he told me. "You need a cast of characters to pull this off."

Bezwoda's study, though, provided a glimmer of hope that the brutal side effects of high-dose chemotherapy were worth the payoff. ASCO had issued a statement saying that together, the trials showed "it is not yet possible to draw definitive conclusions" either way. A team of American scientists traveled to Bezwoda's hospital in South Africa to review the doctor's data and design a new trial following his exact methods. But when the team arrived in Johannesburg, Bezwoda was evasive, limiting the researchers' access to himself and his records. What the American scientists discovered even without Bezwoda's full cooperation was enough. Only one-third of the doctor's patient records for an earlier high-dose chemotherapy study published in 1995 could be located. There were no consent forms on file, a standard part of every clinical trial that proves patients understand and accept the risks of experimental procedures. Most of the patients whose records were found were poor, black South Africans who, in the waning years of apartheid, were more likely than their wealthier, white counterparts to be uneducated and illiterate, incapable of providing informed consent for a trial. Perhaps most damning, the U.S. researchers found that the rules and procedures for Bezwoda's trial had been written years *after* the study began rather than before it was launched. Bezwoda was a fraud, his trial a farce.

Bezwoda was fired from his hospital in Johannesburg.

ASCO released a letter saying the doctor had falsified his data, and American insurance companies began refusing coverage for high-dose chemotherapy for breast cancer patients. For a short while, some U.S. oncologists said they would continue anyway. To Hortobagyi, it seemed these oncologists were putting faith ahead of science. "We all wanted to believe. It's a good thing to want to believe when you go to church, but not in science," Hortobagyi told me.

The discovery of Taxol as well as the high-dose chemotherapy debacle helped breast cancer treatment turn a corner. The medical oncologists who prescribe drug treatments now approach the disease with two questions in mind: Which drug regimen is most likely to cure the patient? And how much toxicity can be avoided?

Taxanes, the class of drugs that include Taxol and the chemotherapy Nancy Lankford was prescribed, can cause hair loss and leave some breast cancer patients with temporary or permanent nerve damage in their fingers and toes. But unlike an older, popular class of chemotherapy called anthracyclines, taxanes rarely cause irreversible life-threatening side effects. Anthracyclines, derived from bacteria, were discovered in the late 1960s through an Italian program similar to the NCI's that searched for anticancer compounds in the natural world. The most popular of these drugs, a medication called doxorubicin, looks bright red in an IV bag or syringe. Patients call it the "red devil." A small percentage of patients who receive anthracyclines for breast cancer end up with irreversible heart damage. In a small percentage of this group, the damage is so severe that patients require heart transplants. Other drugs given in combination with anthracyclines can increase this risk. As a result, the number of women treated with anthracyclines has fallen dramatically in recent years, while the number who receive taxanes has increased. Trials have shown that taxanes are just as effective for most breast cancer patients. Yet there are oncologists who still favor anthracyclines nearly across the board. To understand why these doctors have bucked the trend of favoring less-toxic chemotherapy, I had to go to New York.

7

pick your poison

As I rounded a corner on Second Avenue in Manhattan, I saw a woman in a pink turban. She looked to be in her seventies, but surely the cancer had aged her, so I couldn't be sure. The woman was sobbing, her hand clasped over her mouth and her shoulders heaving with each short breath. A second woman, younger, had an arm around her crying companion. The younger woman looked up at me as I closed the distance between us. She flashed a weak smile that I returned, a brief moment of recognition, not personal but experiential. All the people going in or coming out of Memorial Sloan Kettering's Evelyn H. Lauder Breast Center have something in common.

It was raining, one of those hot New York City summer showers that make everything feel sticky. I closed my umbrella and walked into the marble lobby of the breast center. Plush sofas and chairs were arranged in a few large seating areas lit by golden table lamps. Several of the walls had smooth, floor-to-ceiling wood paneling. Tall, healthy potted ferns stood in the corners of the room. As I walked to the security desk, I noted some words embossed on a wall: THIS LOBBY IS DEDICATED IN HONOR OF LARRY NORTON, M.D. CARING PHYSICIAN, DISTINGUISHED RESEARCHER, VISIONARY MEDICAL LEADER, AND FRIEND. The text assured me that I was in the right place. I had come to Sloan Kettering to interview Norton, arguably the most influential living breast cancer doctor in America.

As I boarded the elevator, manned by a security guard in a suit, I noticed a bench where weary patients could rest during their brief trips between the building's sixteen stories. When the elevator doors opened on my floor, I was met by one of Norton's assistants, who guided me to a large conference room. On the way, I passed a sea of cubicles where breast center employees were scheduling appointments and conducting other business.

The breast center building, a 236,900-square-foot behemoth housing an imaging center and services for Sloan Kettering's breast cancer patients, opened in 2009. The construction project had been partly funded by a fifty-million-dollar gift from Evelyn H. Lauder, promoter of the pink ribbon, founder of the Breast Cancer Research Foundation and Norton's benefactor. Before her death in 2011, Lauder had procured more than four hundred pieces of art that adorn the walls of the breast center. World-famous violinist Itzhak Perlman performed at the grand opening, playing Mendelssohn's String Octet in E-flat major, op. 20.

The predecessor institutions of Memorial Sloan Kettering, perhaps the most well-known cancer hospital in the world, were founded and nurtured by industrialists, including members of the Astor and Rockefeller families, as well as two former General Motors executives. For more than a century, the hospital has been a mecca for patients seeking advanced cancer care, a proving ground for research into new treatments and a premier employer for some of the best minds in cancer science. For decades, Norton's name has been synonymous with breast cancer treatment and research at Sloan Kettering. A New York City native—his father was a journalist at the *New York Post*—Norton went to medical school at Columbia University and worked at several hospitals in the city before landing at Sloan Kettering in 1988, where he has headed the breast cancer program ever since, churning out research and treatment protocols that have helped shape modern breast cancer treatment across the country.

Brilliant and strong-headed, Norton has earned just about every honor that can be bestowed on cancer researchers and has been a part of nearly every breast cancer treatment sea change and debate. With a colleague, he developed a model for tumor growth that bears his name and has influenced chemotherapy for breast cancer patients around the world. The physicians who designed and directed my care at UCLA disagree with Norton on which type of chemotherapy works best for patients like me, an anthracycline-based regimen or one based on a taxane. I had a list of topics to discuss with Norton, but most of all I was eager to ask him about this discrepancy. I had been treated with the taxane docetaxel and a chemotherapy drug called carboplatin, derived from the element platinum. Which treatment a breast cancer patient receives can depend on where she is in the country. Sloan Kettering's reputation and experience reverberate throughout the East Coast, where women are more likely to be treated with an anthracycline-based regimen. On the West Coast, where UCLA is influential, they are more likely to be treated with taxanes.

When the seventy-two-year-old Norton entered our meeting room, I noticed that his hair was whiter than in his hospital headshot. Norton began his career in the 1970s, when chemotherapy was still relatively new. Like Susan Love, Norton was fond of analogies and he had one for the anthracycline/taxane debate. He likened it to different musical styles popular in the mid-twentieth century. "There was East Coast jazz and there was West Coast jazz," said Norton, "and there is East Coast oncology and West Coast oncology. They're very different in their approaches to the disease." To Norton, the different approaches are cultural, not scientific. "The New Yorker will say, 'What's the very best thing for me?' And the person in California will say, 'What's the thing for me that's pretty good but is the least harmful?'" said Norton.

Large-scale trials evaluating anthracyclines and taxanes have found that anthracyclines may work better only for women with a certain subtype of breast cancer and that the difference is small. One trial that included more than three thousand women with this subtype found that in the group of women treated with anthracyclines, docetaxel, and a third drug designed to target the subtype's particular mutation, 84 percent were alive without breast cancer more than five years after treatment. In the group treated with docetaxel, carboplatin, and the targeted drug, 81 percent were alive without breast cancer. Statistically, the difference in efficacy between the regimens was not significant. When it came to heart damage, however, the difference was dramatic. Two percent of women treated with anthracyclines, docetaxel and the targeted drug developed congestive heart failure, five times higher than in the group of women who received docetaxel, carboplatin and the targeted drug. (A third group of women in the study received anthracyclines and docetaxel but not the targeted drug, and they were nearly twice as likely to suffer congestive heart failure.) The targeted drug, designed for women with a particular subtype of cancer, also increases the risk of heart damage. My breast cancer was this kind, which is part of the reason my UCLA oncologists chose a taxane-based regimen for me.

The National Comprehensive Cancer Network, a nonprofit consortium of major U.S. cancer centers that set guidelines for oncologists, endorses both taxane- and anthracycline-based chemotherapy regimens for breast cancer patients, leaving it to oncologists to choose their preferred medicines. At Sloan Kettering, oncologists tend to choose anthracyclines. Lianne Zhang, who was diagnosed with breast cancer in 2016 at the age of thirty-two, told me she consulted with an oncologist at Sloan Kettering who was adamant she undergo treatment with anthracyclines, even after Zhang, who had done her own research, inquired about using a taxane-based regimen without anthracyclines. Turned off by the oncologist's absolutism, Zhang received chemotherapy at Johns Hopkins instead. Her oncologist there prescribed a non-anthracycline taxane-based regimen, the doctor's preferred protocol for breast cancer patients.

Norton told me he favors anthracyclines simply because he believes that they are more effective, even if the difference is slight. "If you're standing in a dark room and you hear a gunshot, the odds are that the bullet is not going to hit you, but does that mean you shouldn't duck? I'm an East Coast doctor," Norton said. "I duck." As Norton explained his reasoning, I realized that the debate over anthracyclines versus taxanes is similar to the consternation over breast cancer screening. As with mammography, there is no clear line. There is only a value judgment. Are you willing to risk a serious side effect for a slightly better chance your cancer will be cured? Or are you willing to risk a slight chance you will not be cured if it means you could avoid a serious side effect? (Anthracycline-favoring doctors point out that the heart damage that results from the chemotherapy can sometimes later be treated with medication.)

Different women might make different choices, but in oncology they rarely get that chance or even know that they have a choice.

My father is not a sentimental man. He doesn't have a shoebox of mementos. He didn't save any old toys from his childhood. I can't recall ever seeing him take a photograph. And yet, for the entire course of my cancer treatment, the home screen on my father's battered iPhone was an image of me standing on Venice Beach. My dad said he liked the picture because I looked so healthy.

He was right. I did look surprisingly good for someone three months into chemotherapy treatment. In the photograph, I'm smiling wide and wearing oversize sunglasses and large silver hoop earrings. My long wavy brown hair cascades over my right shoulder, and a canvas purse strap bisects my torso, which is covered by a distressed gray V-neck T-shirt. I look fit and happy. The Pacific sparkles behind me, and the shadows of palm trees fall over the sand.

The only sign that something is off is a black band around my upper right arm. It looks out of place, a dark rectangle in an otherwise breezy beach snapshot. The stretchy fabric hid the end of a catheter that dived under my skin and ran, inside a vein, up my arm and across my chest, ending at my heart. In all, it took forty centimeters of purple plastic tubing to deliver the chemotherapy drugs into my body every three weeks.

The picture was taken by one of my aunts. She and my uncle had come to Los Angeles from upstate New York to care for me after my fourth of six chemo infusions. I appreciated the effort and the company, but I had spent most of the weeklong visit feeling guilty there wasn't more for them to do. I didn't need my aunt to hold my hair while I threw up because I never vomited thanks to prescription pills and medical marijuana. I didn't need an extra blanket because I never felt a chill. No one needed to drive me to run errands because I drove myself. The day my aunt photographed me on the beach, we were on our way to the airport. I had stopped in Venice to squeeze an afternoon of bright west LA fun into a trip that was mostly spent chatting inside our small house. Two days after my aunt and uncle left California, I headed off on a weekend camping trip with my husband and three-year-old daughter.

Back then, if I didn't tell you I had cancer, you didn't know. If you saw the armband from afar, you might have thought it was a contraption to hold an iPhone while jogging. I chose an arm catheter because I could cover it up and it didn't leave a scar like the more traditional ports that are implanted in the chests of many chemotherapy patients. I paid $2,600 to rent ice-pack headgear that I wore during infusions to minimize the impact on my scalp so my hair didn't fall out. Used in Europe for decades, scalp-freezing is becoming increasingly common in the United States, eliminating a side effect that, for many women, is the worst part of breast cancer treatment. Chemotherapy drugs circulate through a person's entire body during an infusion. Caps chilled to thirty degrees below zero and Velcroed to the patient's head reduce the blood circulation in the scalp and the flow of chemotherapy drugs that damage hair follicles.

During my chemo sessions, Collin Velcroed a new frozen cap to my head every twenty minutes to keep my scalp as cold as possible. We brought an electric blanket to the infusion center to keep my body warm. By the time I stood on the beach in Venice, my hair was thinner but still there. My eyebrows had disappeared, but I hid their absence behind thick, dark-framed glasses. I wasn't ashamed of my cancer diagnosis, but keeping my hair gave me privacy. If I didn't want to talk about my cancer with a stranger or acquaintance, I didn't have to. I could be just another mother picking up her three-year-old girl from day care. I could be just another grocery-store shopper squeezing avocados and ordering sliced cheese at the deli. I could be just another dog owner out for a walk around my neighborhood in northeast Los Angeles.

There were small signs that I was enduring something out of the ordinary, but they were imperceptible to all but the most careful observers. The skin on the backs of my hands had turned pink and begun to peel. Each of my fingernails was marked with faint horizontal white bands. Called Mees' lines, the markings can be caused by exposure to arsenic. I had one line for each time I got chemo. Once, I was having a pleasant outdoor lunch with my brother and we were deep in conversation when I noticed his face suddenly turn pale. "Your nose is bleeding," he said, handing me a napkin. One of the intravenous drugs I received made the skin inside my nostrils paper-thin. It could break down at the slightest sniff.

There were other physical changes known only to me. I had

alternating diarrhea and constipation every day. The fingernails with the white lines felt loose, like they might suddenly break free of my fingers. The tips of my thumbs were numb. Taste-bud changes at first made anything alcoholic taste like metal and progressed until eventually nothing had any taste at all.

Although I felt fortunate to avoid an anthracycline that could have damaged my heart, chemotherapy left me weak and exhausted. There is no chemotherapy regimen that does not wreak at least some havoc on the body. As doctors and scientists have moved away from the most toxic forms of the treatment and toward chemicals with fewer harsh side effects, they have also increasingly gravitated toward an even more attractive option—no chemotherapy at all.

Minneapolis winters are serious business, so when the city's famed Hubert H. Humphrey Metrodome opened its doors to inline-skating enthusiasts in the 1990s, the stadium offered a rare and welcome place to get some exercise on a weekday night. In 1993, Carolyn Streed, a forty-four-year-old insurance marketing representative, had finished some laps around the stadium concourse and was changing out of her skates when a man approached her. He asked about Streed's T-shirt, a souvenir from a fund-raising event she had attended. Streed could tell he was flirting.

The man was middle-aged, like Streed, and handsome. For weeks after, whenever the two saw each other at the Metrodome, they would skate together. Streed was single. She had never gone out of her way to meet men. "I think you just go about life doing the things you love—sooner or later, you'll run into somebody," she told me.

Finally, one night at the Metrodome, the man worked up the nerve to ask for Streed's phone number. Would she like to go out sometime? A few weeks later, Streed and the man, a warehouse manager named Mike, had dinner. Within a month, they were living together. A year later, they got married. The late-morning wedding was held in an Episcopal church on the shores of Lake Minnetonka. Streed organized a luncheon of salmon and chicken for the one hundred or so guests at the reception.

After a few years as husband and wife, the couple sold the home-inspection business they had started together and settled into retirement. Streed had always loved watercolors and wanted to see if she had a knack for painting. She signed up for some classes at a local art school near her home in Hudson, Wisconsin. Within two years, she was selling her paintings, most of them pet portraits and still lifes. She was even teaching others in a studio Mike had built for her over their garage.

But Mike felt restless. He wanted to buy a motor home and travel the country with his wife for as long as they could manage it. "He always supported my dreams. Now it was time for me to support his," said Streed. The couple had a huge garage sale, put in storage the items that remained and rented out their house. "We were on the road two and a half years," said Streed.

They went to national parks in the west and Cape Cod in the east. They spent two summers parked on the shore of Lake Superior. For old times' sake, they even participated in an inline-skating marathon in Duluth. When the couple grew tired of the nomad life, they bought a new house in Tucson. In the mornings, they hiked, exploring the canyons around town. Streed loved the bright sunlight and dark shadows of the high desert and started painting landscapes.

As a woman who had found her two great loves, her husband and art, late in life, Streed felt immensely blessed. Then, one day in mid-2016, she noticed a few spots of blood in her underwear, years after she'd had her last period. She was diagnosed with endometrial cancer, a disorienting bit of awful news in the midst of an otherwise blissful life. Streed soldiered through a hysterectomy and was subsequently declared cancer-free. She was grateful to have survived and relieved that her treatment plan didn't include chemotherapy.

But after the surgery, it seemed to take forever for Streed to recover. First, she developed pneumonia that sent her back to the hospital, and then a mysterious pain led to surgery to remove her gallbladder, which was followed by a buildup of fluid in her lungs. It seemed the surgeon who had removed Streed's gallbladder may have nicked her lung during the surgery. When she finally got back home, she took an oxygen tank with her. That fall, Streed's husband took her on a monthlong RV trip. He hoped it might cheer her up and help put all her health problems in the past, but Streed had a cold for most of the trip and felt weak and exhausted.

As the holiday season approached and Tucson turned cold, Streed finally felt better. She could glimpse her old life again. She was a few organs lighter, but she was hiking and painting. "I was just getting my mojo back," Streed remembered, "when I was hit with the breast cancer diagnosis."

A routine mammogram had turned up a mass in her right breast. "I kind of in a sense knew that it was going to be cancer," said Streed. A biopsy confirmed the mass was malignant. An MRI revealed a second tumor. The lumpectomy Streed was hoping for was out of the question. "I'd just been sucker-punched," she said. Streed had a weekend to absorb the news before she underwent a mastectomy. "Intellectually I knew that my breasts aren't who I am, but I have always been nicely endowed and they were certainly a big part of my womanhood and a huge part of my pleasure." Streed couldn't make her mind up about whether to get reconstructive plastic surgery, so she didn't. The mastectomy had left the right side of her chest flat, with an ugly scar. Streed had learned that she didn't like the way prescription opiates made her feel, so she didn't take them, but the Tylenol she relied on instead at home after the surgery didn't soothe her agonizing pain. What a turn. Eight months earlier, she had been a working artist enjoying retirement with the love of her life. Three surgeries later, she had only one breast and was in pain, a cancer patient twice in one year.

After two weeks recovering at home, Streed felt well enough to drive to her gallery on the east side of Tucson. She hoped returning to her art might get her mind off cancer. On the way home, her phone rang. It was her surgeon. "She asked how I was doing," said Streed. The truth was, not very well. "Will I ever feel joy again?" Streed asked her doctor. "And she said, 'Oh Carolyn, yes, you will. I promise you will. In fact, maybe you'll feel joy right now.'" The doctor had good news.

Cancer patients get used to awkward conversations. Some friends, newly informed of my unfortunate situation, told me stories of family members who had been diagnosed with cancer and died. I understood why they told me these stories—the desire to relate is strong—but they were not encouraging. Others, displaying a tic I came to understand was about reassuring themselves my fate would not befall them, asked if the disease ran in my family. (It didn't.) Some people asked me if I was at least happy that I would end my cancer experience with brand-new breasts. (I was not.) But by far the most common question I fielded from those who found out about my diagnosis was: What stage is it?

For nearly a century, doctors have relied on staging systems to determine the prognosis for various cancers. The first staging system applied to rectal cancer and was developed by a British pathologist named Cuthbert Dukes in 1932. In the 1940s, a French doctor named Pierre Denoix developed a staging system that applied to solid tumors, and in 1958 the International Union Against Cancer published the first guidelines for world-wide use. For solid-tumor cancers such as breast or prostate cancer, staging depended on the size of the primary tumor and whether it had spread to nearby lymph nodes or other parts of the body. Cancers ranged from stage I, for those that were small and self-contained, to stage IV, for those that had traveled through the bloodstream to invade other organs or bones.

A lot has happened since the 1950s. In the case of breast cancer, doctors now know that size doesn't matter as much as previously thought. The *type* of malignancy a woman has is just as important, if not more so, than the width of a breast cancer tumor and whether it has spread to nearby lymph nodes. That question—what stage is it?—has become far more complicated. In 2018, the American Joint Committee on Cancer, which sets staging guidelines for U.S. doctors, dramatically updated its criteria for classifying breast cancer to include estrogen-receptor status and a host of other factors, such as grade, which measures how abnormal cancer cells look under a microscope, and whether or not a woman's cancer has an overabundance of a protein that can fuel its growth. Breast cancer patients increasingly undergo chemotherapy and other drug treatments before surgery. (This happened to me.) These treatments often shrink tumors before they are cut out of the body, meaning there is no way to know precisely how large their original tumors were or how many, if any, lymph nodes contained cancer. These women never know exactly what stage their breast cancer was.

The new breast cancer–staging guidelines also now include a genomic test that evaluates the likelihood that a woman's disease will recur without chemotherapy. For as long as chemotherapy has been available, doctors have known that it does not work for everyone. Some women might be cured of the disease without suffering the toxicity of the drugs; others will see their cancer recur regardless of treatment. But until recently, there was no way to tell one group from another beyond the staging system, which was imprecise but better than nothing. Genomic array tests were developed in the early aughts on the heels of the federal government's three-billion-dollar project to sequence the human genome. Currently, the most widely used genomic test for breast cancer is called Oncotype DX. For women with the most common type of estrogen-receptor-positive breast cancer that has not spread to nearby lymph nodes, Oncotype DX analyzes a tumor sample and provides a score between 0 and 100. The lower the score, the lower the probability a woman's breast cancer will recur if she's treated with surgery and hormone drugs but skips chemotherapy. The first trial to prove the test's efficacy analyzed tumor samples collected between 1982 and 1988 through the National Surgical Adjuvant Breast and

Bowel Project (NSABP), a massive research effort that has for decades allowed scientists and doctors to study many elements of breast cancer care. Because the NSABP trial in those years had tracked its patients and saved their tumor tissue, researchers in 2003 could analyze the tissue with the Oncotype DX test and compare scores against the known fates of the women; they verified that the test accurately predicted who was at high and low risk of recurrence.

The Oncotype DX test has markedly reduced the use of chemotherapy in breast cancer. For the first time, oncologists considering chemotherapy for their patients have another factor to weigh beyond the size and type of cancer present. A high Oncotype DX score can convince a woman reluctant to undergo chemotherapy that the treatment is warranted. On the other end of the spectrum, a low score offers peace of mind to doctors whose patients refuse chemotherapy.

When Carolyn Streed's surgeon called her in the car that day in 2017, she relayed the results of Streed's Oncotype DX test. Streed's tumor scored a 22, putting her in an intermediate-risk group. For a decade, many doctors have not known what to recommend to these women. Some have gotten chemotherapy; others have not. Streed decided to skip chemotherapy. Six months later, a study involving some 6,700 intermediate-risk women with early-stage breast cancer who underwent surgery (and sometimes radiation) and were then randomly assigned to receive hormone treatment and chemotherapy or hormone therapy alone showed that Streed had made the right choice. Although the study found that some of the small percentage of intermediate-scoring women diagnosed with breast cancer before the age of fifty benefited from chemotherapy, most women with intermediate Oncotype DX scores could safely skip the treatment. Ninety-four percent of the women in each group were still alive nine years into the study.

Overall, new knowledge about the value of the Oncotype DX test means that about two-thirds of women who would have been advised to receive chemotherapy based on staging alone can now safely avoid the treatment. This adds up to some sixty thousand women every year.

My experience with chemotherapy was not nearly as brutal as it has been for women in the past. Thanks to scalp-freezing technology and drugs to blunt the chemotherapy's side effects, I worked throughout my treatment. The skin on my hands eventually returned to normal, the numbness in my thumbs went away, and, possibly because I had not been treated with an anthracycline, I emerged without any apparent long-term damage to my heart or other organs. That's not the case for all women. Some former patients I have interviewed were not able to work or care for their children during chemotherapy and suffered long-term side effects, including permanent numbness in their hands.

Chemotherapy, from anthracyclines to taxanes, represented a huge leap forward in treatment for breast cancer patients. Now doctors know that taking a few steps back is also progress.

8

targets

The sunny February day I visited the headquarters of the drug company Genentech, it took me twenty-five minutes to find a parking space. As I weaved through lot after lot, I grew increasingly panicked. I had a nine-thirty appointment with Courtney Aberbach, a young public-relations expert who would escort me around the company's property in south San Francisco. We had a packed schedule and I did not want to be late. Finally, on the sixth floor of an employee parking garage at the top of a hill, I found a spot. As I hurried back down on foot, along a road called DNA Way, I saw San Francisco Bay behind the large buildings on Genentech's sprawling campus. There were fifty-seven buildings in all, housing everything from corporate offices to laboratories.

With about fifteen thousand employees worldwide, Genentech anchors south San Francisco's teeming biotech industry. The company was founded in 1976 by Robert Swanson, a twenty-nine-year-old venture capitalist, and Herbert Boyer, a biochemist at the University of California, San Francisco, who had discovered, in collaboration with Stanford's Stanley N. Cohen, how to splice genetic material from one species with genetic material from other species. Recombinant DNA allowed scientists to modify genes, including those in humans. At Genentech, Boyer and his team deposited recombinant DNA into isolated cells in laboratories. The cells multiplied, cloning the modified genetic material, which could be made into medicine. Called "biologics," these medicines opened a new frontier in drug development.

Genentech went public in 1980, two years before the pharmaceutical company Eli Lilly licensed Genentech's first market-ready product, a synthetic insulin cloned inside benign *E. coli* bacteria. Before Genentech's drug hit the market, diabetics had relied on insulin harvested from the pancreases of animals. Eli Lilly needed more than fifty million animals per year to produce enough medicine for its U.S. customers. A drug made purely in a lab provided an unlimited supply. After finding a way to produce synthetic insulin, the company cloned a blood-clotting protein for hemophiliacs, a growth hormone for children and a blood thinner for heart attack patients. The last, named Activase, was a huge success. When it hit the market in 1987, the drug put Genentech at the forefront of America's biotechnology industry. Swiss pharmaceutical company Roche bought a majority stake in Genentech in 1990, right around the time the biotech firm was working on a new class of drugs to treat cancer. The effort eventually transformed Roche into one of the world's top makers of cancer drugs.

I had come to Genentech headquarters to see the birthplace of one of the most successful of these drugs, a breast cancer medicine, and to observe what profits from that drug and others had built—namely, the company's ever-growing campus, so crowded with employees that parking spots were constantly in short supply. "Yeah, I forgot to tell you that parking here can be tricky," Aberbach said when we met in the lobby of one of Genentech's modern glass-and-gray-stone buildings. I was still sweating from my hurried walk as Aberbach handed me a temporary sticky ID badge and we set off for Building 51, a packaging facility where biologics made at other Genentech manufacturing plants are processed and shipped across the United States and to other countries. Aberbach had arranged for me to get a look inside.

First, though, we had to strip. Aberbach had warned me that the plant was a sterile environment and the rules for entering the facility were strict. No makeup, no jewelry, no nail polish. Clear lip balm was allowed, but I would be prohibited entry if I had had contact with a rodent within the previous twenty-four hours. Sufficiently devoid of potential contaminants, Aberbach and I entered a locker room inside the plant with Carrie Schimizzi and April Maguire, Genentech employees who would lead the tour. "You have to take everything off but your underwear," Schimizzi told me.

Aberbach and I, previously unaware of this requirement, exchanged a nervous look and turned our backs to each other as we disrobed and stuffed our clothes into lockers. Schimizzi, also in her underwear and bra, led us into a second, smaller changing room and handed me a plastic-wrapped package containing a navy-blue cotton jumpsuit, which I pulled on. I noticed a line painted on the floor as Schimizzi passed me two disposable shoe covers. "Put one on and step over the line with that foot. Then, you put the other one on and step over with your other foot," she said. It was an awkward procedure meant to ensure no outside dirt could be tracked inside the plant. We proceeded through a third door into yet another room stocked with more jumpsuits. These were a lighter blue and would go over the first suit. Schimizzi explained how to unfold the garments and put them on without letting the fabric touch the floor, flinging her own suit open and then flipping it halfway inside out so she could step into it. This was Schimizzi's third time suiting up that day. To our outfits, we added safety googles, hairnets and gloves that were secured over the cuffs of our suits. We looked ready to enter a hazmat site. In the hallway, we met a third plant employee, Shiraz Ali, who was wearing a paper covering over his beard.

The two-story facility was surprisingly compact, with a central core of small rooms and a wide hallway wrapping around the outside. Inside the central rooms, which had large windows, I could see various pieces of assembly line equipment. One room was devoted to a machine that could wash and sterilize one hundred glass vials per minute. In another, Schimizzi pointed out three-hundred-liter stainless-steel tanks that looked like small spaceships, with hatches and tubing running from the tops. One room housed a contraption called a depyrogenation tunnel, which heats drugs to 330 degrees Celsius in a type of super-sterilization. Another room contained a machine called a lyophilizer. Schimizzi told me that the machine turned liquid medicines into freeze-dried white powder, which was easier to ship than liquids and more shelf-stable. (Once the medicines arrived at health-care facilities, the powder would be rehydrated.) Throughout the plant, I saw masked workers wearing blue and white suits and holding clipboards. They documented parts of the drug-processing procedure that ensured Genentech products were clean and safe before they left the facility.

Watching workers inside the packaging plant, I saw that the steps required to get a drug from a manufacturing facility to a packaging facility and finally to a hospital were highly choreographed. The stages were organized, standardized and predictable. But the process that came before the one I saw on the company's campus—the invention of Genentech's cancer drugs—well, that was another story entirely.

In the mid-1970s, right around the time Robert Swanson and Herbert Boyer were starting Genentech on the heels of recombinant-DNA research, two scientists at the University of California, San Francisco, made a discovery that would reset the field of cancer research and pave the way for Genentech to grow into the behemoth it is today. UCSF microbiologist Michael J. Bishop and a postdoctoral fellow named Harold Varmus were looking for clues about cancer's genetic origins by studying a cancer-causing chicken virus. The conventional wisdom among scientists was that genes inside such a virus could cause normal cells to mutate and divide uncontrollably. But Varmus and Bishop, in a finding that would earn them the Nobel Prize, proved that the chicken virus actually transferred a specific normal chicken gene from one cell to another. The transplanted gene, called an oncogene, controlled growth, and when it was moved to a new location, it could trigger uncontrolled cell division. Varmus and Bishop's major discovery was that all healthy animals, from chickens to humans, possess genes that, when triggered, can cause cancer. Molecular biologists around the world raced to pinpoint other oncogenes. If they could be identified, the thinking went, they could be targeted and defeated.

One particular oncogene called *HER2* drew the interest of a Genentech scientist named Axel Ullrich. Normally, *HER2* controls the growth and repair of breast cells. But working with scientists from England and Israel, Ullrich helped prove that a mutated version of *HER2* could cause uncontrollable cell growth—that is, cancer.

A few years after making this connection, Ullrich had a chance meeting with a tall, mustachioed UCLA oncologist named Dennis Slamon. The son of a coal miner who was the son of a Syrian immigrant, Slamon had been the first in his family to go to college. He enrolled in Washington and Jefferson College, an hour-and-a-half drive from his hometown of New Castle, Pennsylvania. He paid his tuition with a scholarship and money he earned working summers in a mill that processed steel used to make axles and springs for tractor-trailers. After he graduated, Slamon went to medical school at the University of Chicago, earning a PhD in cell biology as well as his MD. He finished his studies right around the time Bishop and Varmus discovered oncogenes and made cancer the hottest field to work in for a medical researcher.

Slamon had spent his entire career at UCLA. He took care of cancer patients but he also conducted research and collected tumor samples he could study in the lab. Slamon investigated oncogenes in mice and later studied oncogene activity in various types of human cancer, searching for genetic mutations that might one day be targeted with drugs. Other than tamoxifen, treatment for breast cancer at the time was indiscriminate. All women diagnosed with the disease underwent surgery, chemotherapy and radiation, or some combination of these treatments. Sometimes, the treatments worked. Too often, they did not. Like many researchers around the country, Slamon was on the hunt for a genetic target that caused cancer and could be defeated with a drug.

When Slamon met Ullrich—they bumped into each other at an airport after a 1986 scientific conference in Denver and later had a long dinner in Los Angeles—the two got to talking about their work. Slamon had his cancer-cell samples at UCLA and Ullrich had Genentech's sophisticated technology that could be used to conduct gene tests on the samples. They decided to partner up and investigate whether any of Slamon's tissue samples contained a genetic target directly related to cancer. Before long, they got a match. Ullrich's *HER2* gene probe appeared to match a genetic component present in some of Slamon's breast cancer cells.

To know for sure, Slamon enlisted the help of Bill McGuire, an oncologist at the University of Texas who had amassed a much larger bank of breast cancer tumor samples along with detailed information on the patients from whom the samples had been retrieved. When the scientists searched the bank, they found that *HER2* was increased and the protein that the gene made was overexpressed in a sizable fraction of the samples, with many more times the amount of HER2 proteins than typical breast cancer cells have. In these particular cells, the growth-factor signal of *HER2* had been sent into overdrive, causing tumors to grow. In a second critical finding, the scientists realized that the tumors with overexpressed HER2 proteins had come from patients whose cancers were aggressive and tended to recur in deadly fashion after initial treatment. For decades, oncologists had used the traditional staging system for cancer, trying to predict which breast cancers were most deadly by measuring the size of tumors and whether they had spread to a woman's lymph nodes. But *HER2* was something else, a genetic mutation that correlated with a poor prognosis, often regardless of tumor size, lymph nodes or response to traditional treatments.

Slamon, Ullrich, McGuire and several other researchers involved in the *HER2* research published their work in the prestigious journal *Science* in 1987, but the larger world of oncology research was slow to react to their discoveries. Some researchers reported that they struggled to reproduce the results documented in the *Science* article. Others pointed out that just because some breast cancer tumors overexpressed HER2 proteins did not necessarily mean that the genetic mutation was the element fueling the cancer. More research was needed before a targeted drug could be developed, so Slamon, who had been studying various genetic mutations in many types of cancer, got to work with a laser-like focus on breast cancer. With some collaborators, he proved that it was possible to genetically alter cells to overexpress HER2 in the lab. Then Slamon implanted human HER2-positive tumors in mice and watched as the masses grew ferociously and spread quickly. Genentech scientists, meanwhile, set to work inventing a compound that could stymie the action of the protein. Using recombinant-DNA technology, the company's scientists coaxed ovary cells from a rodent known as a Chinese hamster to develop an antibody to the mutated HER2 protein. In petri dishes and mice, the antibody was miraculous, halting cell division and eventually causing cancer cells to die. Slamon was eager to start human clinical trials to see if the antibody performed just as well in women with HER2-positive breast cancer.

The corporate leaders at Genentech, however, were not as excited by the progress as Slamon, Ullrich and others working on the HER2 research were. The HER2 antibody was a difficult and expensive drug to produce, plus a string of cancer-drug failures within and outside the company had left Genentech executives gun-shy about new experimental cancer medicines. Disappointed, Ullrich left the company and returned to his native Germany to head up a scientific institute in Munich. But Slamon persisted, repeatedly flying up to San Francisco to personally lobby Genentech executives to continue funding HER2 research. He was, by all accounts, relentless.

Slamon also secured a series of grants from makeup giant Revlon to ensure his research could continue. After Roche acquired its stake in Genentech in 1990, injecting capital into the company, Slamon and the Genentech scientists working on HER2 finally prevailed. The company would finance the first human clinical trials of its HER2 antibody. Like all phase 1 clinical trials, the studies would not primarily test whether the antibody killed cancer; it was still too early in the research process for that. The goal was to see if the antibody was safe. In 1991, Slamon administered infusions of the mouse antibody to twenty women with late-stage HER2-positive cancer. The risk of something going terribly wrong was significant. It was possible that the women's bodies could mount dangerous, and even deadly, reactions to antibodies produced by another species. Luckily, this didn't happen, paving the way for the next phase of research, testing the safety of a new version of the antibody Genentech had since created that contained far less mouse DNA. The "humanized" drug would be administered along with a chemotherapy agent to late-stage HER2-positive breast cancer patients at UCLA, Memorial Sloan Kettering in New York and the University of California, San Francisco.

At UCLA, fifteen women with late-stage metastatic breast cancer received the drug along with traditional chemotherapy. One patient, a woman named Barbara Bradfield, had been treated for breast cancer in 1990. When her cancer reappeared two years later—forming a lump in her neck and lesions in her lungs—she was despondent. She had nearly given up hope for a cure and planned to abandon traditional oncology in favor of unconventional, experimental treatment in Tijuana, Mexico. But Bradfield's oncologist had sent Slamon a sample of her tumor, which tested positive for the mutated *HER2* gene. Slamon called Bradfield himself. Would she give the drug regimen a try? Bradfield declined, but Slamon's persistence had already paid off with Genentech and he was determined to keep up the effort. He called Bradfield back the next day and asked again. This time he convinced her. With Bradfield on the drug combination, the tumor on her neck shrank and shrank. The change was so visible that even other patients in the trial noticed it. By the end of the three-month study, the cancerous lesions in Bradfield's neck and lungs had disappeared completely. Tumors in three other women in the UCLA study group also shrank when treated with Genentech's new, targeted drug, Herceptin. For the other women in the study, the Herceptin-chemotherapy combination did not appear to work.

The next phase of trials took place the following year and produced similarly encouraging results. A significant portion of late-stage breast cancer patients enrolled in the study responded well, their cancers shrinking when hit with the anti-HER2 medicine and chemotherapy. The final step, the one that would determine whether Genentech's drug could be marketed and sold as a cancer therapy, was a large study with multiple trial arms. Launched in 1995, the study included hundreds of patients at sites around the world. The cost of such trials can exceed one hundred million dollars.

As the research got under way, word began to spread of a Genentech wonder drug that could effectively treat some of the most hopeless cases of breast cancer. Women dying of the disease began demanding to be tested for the *HER2* gene and insisting that Genentech make its new drug available to women outside the study. Genentech, furiously trying to make enough of the complex drug for research patients, initially said no, but the company set up a compassionate-use program after patients protested on the company's Bay Area campus and enlisted the support of influential breast cancer advocates.

The study itself had gotten off to a shaky start. In one arm of the study, women with newly diagnosed metastatic HER2-positive breast cancer received chemotherapy and the Genentech drug or chemotherapy and a placebo. Neither the patients nor their oncologists would know which regimen the women were getting, ensuring that responses would be documented accurately. But the uncertainty had a cost. Not knowing whether patients would receive an exciting new drug or a simple saline solution, women and their doctors hesitated to participate. Eventually, Genentech redesigned this arm of the trial. If a trial patient's tumors did not shrink or continued to grow on the study, the company would allow the patient and her oncologist to know whether or not she had been receiving the antibody. The information, though, was kept secret from radiologists measuring the size of tumors. The advocates who had brokered a deal between Genentech and breast cancer patients outside the study also stepped up, helping the company recruit patients to the research effort.

Slamon, observing highly encouraging results among women in the part of the study he headed at UCLA, grew more and more excited as the trials progressed. He personally lobbied the Food and Drug Administration to consider granting approval before the entire research project was complete. In the end, the drug's efficacy was clear much earlier than anyone anticipated. Because HER2-positive breast cancer is highly aggressive and fast-growing, it tends to recur earlier than more common types of the disease that are fueled by estrogen. This meant that it didn't take very long to see if study participants' disease recurred when compared against women not getting the new drug.

In the fall of 1997, a Genentech executive flew to Southern California and met with Slamon in a hotel bar to share the results that the company would present to the FDA months later. The antibody definitely worked. Of the women who received the drug along with chemotherapy, 49 percent saw their tumors shrink by half or disappear entirely; of the women who received chemotherapy alone, only 32 percent saw their tumors shrink significantly. Genentech announced the results of the study at an American Society for Clinical Oncology conference in May 1998, the year before high-dose chemotherapy studies were presented at the same meeting. The FDA soon approved the drug in combination with traditional chemotherapy for patients with metastatic breast cancer. A chance meeting in a Denver airport between Axel Ullrich and Dennis Slamon had been the catalyst for one of the most important breakthroughs in cancer research. Along with tamoxifen, Genentech's HER2 drug, Herceptin, opened a new era in the fight against breast cancer—one in which treatments are tailored according to information embedded in tumor cells.

Genentech, which wanted to expand the market for the drug, soon launched more clinical trials to test whether Herceptin could eradicate cancer in women with early-stage breast cancer that had not yet spread. It could. The FDA expanded its approval in 2006 for all women diagnosed with HER2-positive breast cancer. In 2010, the agency approved Herceptin for use in treating gastric cancer as well.

In the years since, Genentech has become one of the world's leading biotechnology companies, largely due to its success with cancer drugs. In 2017, sales of Herceptin topped seven billion dollars worldwide. The drug has transformed HER2-positive breast cancer from one of the deadliest forms of the disease to one of the most survivable. Some two million women worldwide have been treated with Herceptin, including me.

There were so many ways that luck fell on my side as a breast cancer patient. The most important was that I had HER2-positive disease. My cancer also tested negative for the estrogen receptor, another lucky break. Breast cancers that are HER2-positive and estrogen-receptor-negative tend to have the most durable and dramatic responses to Herceptin. Doctors treating women with breast cancer that is HER2-positive and estrogen-receptor-positive must, in essence, fight the disease on two fronts, which is harder and more complicated. My third turn of good luck was that I lived in Los Angeles and could be treated at UCLA, where Slamon still heads the cancer center's research program. He has trained a new generation of oncologists to follow in his footsteps.

One of his protégées, Sara Hurvitz, is director of the breast oncology program at UCLA and was the quarterback of my cancer care. As a new doctor, she had wanted to treat and study lymphoma patients, but Slamon had persuaded her to specialize in breast cancer. As a smart doctor with multiple publications to her name, Hurvitz is in high demand. Patients who cold-call her office are typically told she does not accept new patients. She makes exceptions, though, if she is recruiting for a clinical trial. In yet another turn of good fortune, Hurvitz happened to be launching a trial sponsored by Genentech in December 2014 when I telephoned to schedule an appointment.

After we toured Genentech's packaging plant, Aberbach and I had lunch in the cafeteria and then walked across the company's campus. We arrived at an office with a hot-pink poster on its door. On the poster, a cartoon woman with purple eyes flexed her upper arm, which was covered by a heart tattoo and a label for Kadcyla, a new anti-HER2 drug Genentech scientists had developed. Called an "antibody-drug conjugate," the medication chemically combines Herceptin with a powerful chemotherapy agent and delivers the compound to HER2-positive breast cancer cells only, sparing healthy ones and causing fewer side effects than standard chemo and Herceptin given separately. On Kadcyla, your hair doesn't fall out.

The office with the poster-covered door belonged to Gail Phillips, a Genentech cell biologist who has spent her entire career obsessed with HER2-positive breast cancer. "It's what I live and breathe," Phillips told me. She had worked on the development of Kadcyla. "What a view," I said as I entered Phillips's office and we shook hands. Through her window, I could see across the roofs of Genentech buildings all the way to San Francisco Bay.

Phillips wore a silver cross around her neck; her long, straight blond hair fell over her shoulders. Her eyes were such a pale green I could practically see through them. Phillips went to college in Texas and she isn't shy about it. On her desk, she had a HOOK 'EM HORNS water bottle and a University of Texas mug that held a collection of pens, and she wore black leather boots with fringe that swung back and forth when she walked.

Phillips is a rarity among the senior scientists at Genentech. She does not have a doctorate in biology or chemistry or any subject. She started a PhD program at the University of Texas Southwestern Medical Center in Dallas but quit before she finished. Despite her lack of a doctorate, after her studies Phillips had packed her clothes and bicycle into her car and, as she said, "I just headed west." It was 1985, five years after Genentech went public, and the Bay Area's biotech industry was gaining steam. Genentech was Phillips's second job interview and she was thrilled when the company hired her to work in a lab.

Self-conscious about not having a doctorate, she spent long, late hours over her microscope. Genentech was a small company back then, just four buildings. By the end of the following year, when Ullrich and Slamon did their important early work on *HER2*, there were nine.

Phillips was part of the team that developed Herceptin. It was a stunning achievement, of course, but it was especially remarkable for Phillips, who began working on the project as a scientist in her late twenties. Many so-called laboratory bench scientists like Phillips never get to help develop a useful, marketable new drug. "In cancer research, you can come up with a lot of cool ideas and develop drugs that look like they work really well. Then you have to do all these rigorous safety studies. Boy, then all of a sudden it's like, 'Wow, this drug is really toxic.'" There are far more failures than successes in the drug-development process. Finding a compound that works in the lab *and* in patients is incredibly rare.

Over the years, as HER2 antibodies became central to Genentech's mission and bottom line, Phillips moved up in the ranks. Today, she has her own lab on the company campus. I was eager to see it, so we left Phillips's office and walked down a hallway, past a tall shelf holding glass beakers and other laboratory accoutrements. Inside the lab, Phillips introduced me to Ginny Li, a principal research associate who has worked for Phillips for fifteen years. "I'd love to show you some of the cells I'm currently growing," Li said. She opened a minifridge under a counter and pulled out a clear plastic container. The vessel, called a flask, was shaped like a flat-sided bottle. Li slid the flask under a microscope and invited me to take a look. The cells were young

HER2-positive breast cancer cells. Through the microscope, they looked like tiny dots in a yellow solution.

Li pulled out another container holding HER2-positive cells that she had been growing for more than a year and treating with Kadcyla. After each addition of Kadcyla, some cells died, but others survived and were treated over and over again. Eventually, Li told me, the breast cancer cells would become immune to Kadcyla, and then they would be studied to understand why, with the hope that a resistance mechanism inside the cells' genes could be targeted with new drugs. Taking care of the cell lines was a delicate job. They had to be fed and cared for gently. Li told me she checked on them every day. "Even on Christmas," she said.

When Genentech scientists were developing Kadcyla, Phillips and other scientists at the company experimented with using different Herceptin-chemotherapy combinations, testing which pairings best found and killed HER2-positive breast cancer cells. One experiment used a highly toxic chemotherapy drug with a fragment of the antibody. It homed in on the cancer cells in petri dishes and killed them quickly and with precision but didn't work so well on animals. Scientists would inject rats with the drug and find them all dead the next day.

Drug development is about efficacy and safety, but it's just as related to profitability, of course. The more patients who can be treated with a drug, the more money a pharmaceutical company earns. Mark Sliwkowski, a biochemist who worked for Genentech from 1991 until 2016, told me that Genentech executives initially hesitated to fully fund the Herceptin project, in part because the market for a HER2 drug, if it could be invented, would be limited to women whose tumors overexpressed the protein. This accounts for only about 15 to 20 percent of all breast cancer patients. But following the success of Herceptin, Genentech decided to go all in on HER2.

In 2012, the FDA approved the company's second HER2-targeted drug. Called Perjeta, the drug functions similarly to Herceptin, but its antibody binds to a different part of the HER2 protein. Clinical trials had shown that Perjeta helped slow tumor growth in women with metastatic HER2-positive breast cancer. Genentech was eager to see if it might improve outcomes for early-stage patients as well. Because of Herceptin, many women with HER2-positive breast cancer never see their cancer recur, meaning there are fewer HER2-positive metastatic patients than ever. Expanding the market for Perjeta, and possibly saving the lives of women who were not cured by Herceptin, was a natural priority for the drug's maker. Even before Perjeta was approved by the FDA for use in patients with metastases, Genentech began a clinical trial to test whether the drug worked for the larger population of women who had early-stage disease.

The trial included about four hundred women and showed that adding Perjeta to a regimen that already included Herceptin and chemotherapy improved the chances that a woman would live for three years without her disease recurring. But a subsequent larger clinical trial involving forty-eight hundred women was a disappointment. Early-stage patients on Perjeta had an 18 percent lower recurrence rate, but that seemingly significant improvement was far less dramatic when one looked closely at the numbers. Four years after receiving Perjeta, 94.1

percent of women were free of invasive breast cancer; in the group that had not received Perjeta, the percentage was 93.2.

Kadcyla, Genentech's third-generation HER2 drug, offered a chance to treat all women with HER2-positive breast cancer with something new. A large trial published in 2012 that led to FDA approval of the drug in 2013 showed the drug increased overall survival in women with metastatic breast cancer. Eyeing the 2019 expiration of Herceptin's U.S. patent, after which other companies could legally manufacture their own similar anti-HER2 antibodies, Genentech wanted to know if Kadcyla might replace Herceptin as the preferred treatment for women with early-stage disease. The prospect of being cured without enduring the side effects of traditional chemotherapy was appealing enough to entice more than four hundred women to enroll in the study, which included sixty-eight sites in North America, Europe and Asia. In the trial, women with early-stage HER2-positive breast cancer would be randomly assigned to receive Kadcyla with Perjeta or the highly effective treatment protocol of traditional chemotherapy with Herceptin, plus Perjeta.

Hurvitz, my medical oncologist at UCLA, was the principal investigator for the Kadcyla trial. She explained the details of the study after I wrangled an appointment with her in December 2014. I had met with several medical oncologists already, trying to find the best doctor to direct my care, but I liked Hurvitz right away. She was a straight-talker. Statistics and data tumbled out of her mouth as quickly as she hopped up onto the exam table and crossed her legs while we talked. Still, I wasn't sure I wanted to sign up for the Kadcyla trial. Hurvitz said she would treat me regardless, but she urged me to consider it. For weeks, as I underwent IVF and prepared to plunge into cancer treatment, I was racked with indecision. As a health-care reporter, I knew how important it was for patients to volunteer for clinical trials. Less than 5 percent of patients diagnosed with cancer in the United States participate in studies, which hinders research. If women had not volunteered to participate in the early Herceptin clinical trials, anti-HER2 drugs would not exist and my odds of surviving breast cancer would have been slim.

But I was worried about being part of an experiment. I was not a metastatic breast cancer patient with terminal cancer and few good options. Existing scientific data predicted that my chances of long-term survival were higher than 90 percent if I underwent standard treatment. In the end, I decided to enroll in the trial, knowing that if I was randomized to the experimental Kadcyla arm of the study, I could always pull out. (Unlike a traditional double-blind clinical trial, in which neither patients nor their doctors know which medication is given, the Kadcyla study was open label, meaning doctors and patients both knew what treatment was provided.) When Hurvitz's study coordinator called to tell me I was assigned to the control group—meaning I could participate in clinical research and still get proven, standard therapy—I was thrilled.

Sheri Weitz, the first woman to enroll in the Kadcyla trial, was hoping for just the opposite. "I didn't want to lose my hair, which sounds very vain," Weitz told me when we talked in January 2018 in the small bungalow near Los Angeles International Airport that she shares with her husband and their son. Weitz has long, wavy red hair that she wears down most days.

"One of the things I would get compliments about was my hair, so, yeah, that was a big deal," she said. When Weitz was diagnosed with HER2-positive breast cancer in mid-2014, she had two malignant tumors in her breast; the cancer had also spread to several lymph nodes in her armpit. Weitz is a registered dietician and had long eaten a strict lean diet and exercised frequently, teaching dance aerobics classes several times a week. *How could this have happened?* she wondered. She had done all the right things, the things everyone says helped prevent cancer. "People would say to me, 'Well, don't you eat healthy? How come you got breast cancer?'" Eat healthy and be healthy is a nice idea, but it isn't a guarantee.

Weitz was randomized to the Kadcyla group. (Eleonora Ford, whose breast cancer was estrogen-receptor- and HER2-positive, also participated in the Kadcyla trial and was in the same arm of the study as Weitz.) After Weitz's first treatment, she swore she could feel the drug working. "I know I wasn't imagining it. I literally could feel these explosions going on where the tumors were," she told me. An ultrasound a few weeks later confirmed what Weitz said she already knew. Her overall tumor volume had shrunk by 25 percent. "Every once in a while, I'd be like, 'Oh my God, I'm a guinea pig,'" Weitz said. Like most women, Weitz had seen depictions of breast cancer treatment on television and in movies in which women were deathly ill. "That was my fear. I'd be a balding woman that would be vomiting. Losing my looks, losing my health, losing my strength," Weitz said. Instead, Weitz told me she worked part-time while in treatment and even taught a few dance classes.

As part of her treatment, Weitz underwent surgery to have her cancerous breast removed along with twenty-one lymph nodes in her armpit. After that, she had five weeks of radiation treatment. Soon after her treatment ended, Weitz started a new job at a local hospital. *Cure* is a tricky word when it comes to an unpredictable disease like cancer, but Weitz is, according to all measurable signs, cancer-free. A second chance at life that seems more precious than ever can embolden a person. The week before we met, Weitz and her husband went on a helicopter ride in Hawaii. "I would have been too afraid in the past, but I just did it," Weitz told me.

In June 2016, my daughter, Evie, graduated from preschool in an elaborate ceremony. The teachers at her pre-K had dressed Evie and her four-year-old classmates in tiny graduation gowns and mortarboards before they walked across an outdoor stage to "Pomp and Circumstance." The old me would have scoffed at such an over-the-top display, but, post-cancer, I relished every milestone. I cheered and clapped during the ceremony and took more than a hundred pictures. Afterward, I headed to the airport to catch a red-eye to Chicago, where Hurvitz would be presenting the results of our Kadcyla trial at the annual American Society of Clinical Oncology meeting, the same conference at which Genentech had announced the results of the Herceptin study in 1998.

I got to my hotel in Chicago the next morning, hoping to check in early so I could take a nap before attending my first conference session. I sat in the hotel lobby waiting for my room to be ready, and the space was swarming with conference attendees. The largest and most important international cancer meeting of the year, ASCO's conference attracts more than thirty thousand people, with more attendees specializing in breast cancer than any other type. Later, as I wandered in and out of clinical trial presentations and panels, I heard the familiar terms cancer researchers use to describe their work. *Stable disease* was cancer that didn't go away but didn't grow either. *Poor outcomes* were patients who died or did not respond to treatment. *Therapy* was a soothing word for pills or intravenous drugs that could cause brutal or deadly side effects. At one panel, I heard about the pain and vaginal dryness that can be caused by the kind of aromatase inhibitors Eleonora Ford takes every day. At another, I learned about efforts to create molecules that could cross the blood-brain barrier, which prevents many drugs from reaching cancerous tumors growing deep inside the brain.

On the third day of the conference, an hour before Hurvitz's presentation about Kadcyla, I walked outside the McCormick Place conference center, which is on the shore of Lake Michigan. I found a shady spot on a concrete wall, sat down and took off the dress shoes that had been pinching my feet all day. To my left, I could see a marina and the downtown Chicago skyline. To my right, the expanse of the lake glittered in the noontime sun. I could hear the water lapping against the concrete wall. The sound mixed with the chirps of birds and the sounds of gears shifting on the bicycles zipping along a path behind me. It was seventy-five degrees and the sky was a soft blue with wisps of high-altitude clouds in the distance.

In the brightness of the Midwestern early-summer day, my mind wandered back to the darkness of the Southern California winter nights after my diagnosis, when I had been the most scared. I flashed to Evie working on her ballerina puzzle and

Collin stepping out the front door to update my father on my diagnosis. My eyes welled up with tears behind my sunglasses. It had been eighteen months since I was diagnosed with breast cancer, but some things hadn't changed at all. Accessing the fear and panic I'd felt on those dark California nights was easy. Would I ever go back to being the kind of person who never cried? From my perch on the wall, I watched a sailboat with a bare mast glide through the channel in front of me. As it rounded the corner into the open lake, the boat's small crew hoisted the sails and cut the engine. The vessel bobbed softly in the cool water until wind filled the canvases and pushed the boat away from shore.

For most of the cancer conference sessions, I sat in the back row so I could duck out easily. Multiple sessions took place simultaneously and I wanted to be able to float from room to room. But for Hurvitz's presentation, I took a seat near the front. The hall was cavernous, with attendees packed into straight rows of seats below six jumbo screens suspended from the ceiling that would project Hurvitz and her PowerPoint presentation throughout the room. The stage up front was brightly illuminated, but the seating area was dim. I looked around for Hurvitz, but in the near darkness it was hard to identify anyone. I wondered if I was the only breast cancer patient in the room or if there were other women from the trials being discussed who had trekked to Chicago to hear the results in person.

Finally, Hurvitz stepped to the podium to deliver her presentation. She was impeccably dressed, as always, wearing a fitted black sheath and an oversize white statement necklace that covered her upper chest. Methodically, Hurvitz explained the trial protocols. One arm, the one I was in, received standard chemotherapy plus Herceptin and Perjeta. The other, Weitz's and Ford's arm, received Kadcyla plus Perjeta. After six rounds of treatment spaced three weeks apart, all women in the trial underwent lumpectomies or mastectomies. Breast tissue and lymph nodes removed in surgery were studied, and pathologists determined whether any cancer remained. After surgery, women in my arm of the trial underwent twelve more rounds of Herceptin and Perjeta, while Weitz and her cohort underwent twelve additional rounds of Kadcyla and Perjeta. Hurvitz explained that 44 percent of women in the Kadcyla group achieved pathologic complete response (pCR), meaning no cancer cells were found in the breast tissue and lymph nodes removed in surgery. (Weitz was one of those women.) Studies have shown that women with HER2-positive, estrogen-receptor-negative disease who achieve pCR are more likely to survive and avoid a recurrence of cancer. The 44 percent figure was impressive. Nearly half of the women in the Kadcyla arm appeared to be cured of HER2-positive breast cancer without traditional chemotherapy. They fared better during the trial and were less likely to discontinue treatment due to side effects.

But as Hurvitz went on, it became clear that Kadcyla would not be replacing standard treatment anytime soon. A larger share of women in my arm of the trial, 56 percent, achieved pCR. I had lucked out again.

How do you greet a person who saved your life? This was the thought that swirled through my brain as I navigated the streets of Los Angeles on the way to Dennis Slamon's office in the fall of 2017. Slamon is well known among the legions of oncologists working in every major cancer center in America and has had more success in the lab than almost everyone. He and Larry Norton, Memorial Sloan Kettering's resident breast cancer guru, know each other well and have clashed and collaborated on research and treatment guidelines over the years. Slamon was the lead author of the large trial showing that anthracyclines given in combination with a taxane or a taxane plus Herceptin offered little additional benefit but a greater risk of heart damage. I thought of Norton's name on the lobby wall of Sloan Kettering's breast cancer center and wondered if Slamon's environs might be similarly posh.

When I walked up to the building that houses Slamon's office and lab, I thought I might be in the wrong place. Before me stood a three-story blue-and-yellow stucco building on a busy stretch of Santa Monica Boulevard in a nondescript part of West Los Angeles. The structure was next door to a Comfort Inn that was AAA-recommended and fronted by faded green awnings. Across the street was a strip mall housing an aquarium store, a Chinese food takeout restaurant and a dry cleaner, among other businesses.

I walked into the UCLA research building's atrium, where an aging, geometric water feature gurgled next to some plastic tables with umbrellas. There were puddles under the tables from a recent rain, and plastic yellow WET FLOOR signs had been set up to warn passersby. Before heading to Slamon's office, I ducked into a bathroom just off the atrium and past some cardboard duct-taped to the tile floor. In the bathroom stall, a sign told me to jiggle the handle after flushing so the toilet wouldn't keep running. The button to summon the elevator stayed unlit when I pressed it, but the doors opened a few minutes later and I went up to the second floor and into a small waiting room.

There was no receptionist, so I sat down on a black leather sofa. One wall of the waiting room was made of glass block that had two faded covers of the journal *Science* embedded in it. One was the 1987 issue announcing that the *HER2* oncogene found in some breast cancer tumors strongly correlated with a worse prognosis for patients. The second was a 1989 issue reporting that the *HER2* gene in breast and ovarian cancers was likely related to the behavior of those cancers. Slamon was the lead author on both studies.

I waited alone for about fifteen minutes and then e-mailed Slamon's assistant, whom he shares with several other researchers. She eventually emerged and led me down a hallway to Slamon's office. The first thing I noticed about the space was that it was a little cluttered. The second thing I noticed was that it was hot. The air-conditioning wasn't working that day. "Yeah, something's wrong," said Slamon.

Sitting behind his desk, Slamon was wearing his usual work outfit, a striped oxford shirt with a button-down collar and a nondescript tie. The pairing looked like it could have been purchased at JC Penney or another midrange department store. I knew that Slamon was from western Pennsylvania, and as I took a seat, I mentioned that my mother was from the same region. "You can pick it up in my accent," he said. "We have our own dialect." As Slamon talked, I heard the distinctive drawl of the Rust Belt. A *clinical trial* was a "clinical trawl." *There* was "derr." *Dialect* was "dialeck."

I asked how long his office had been in the blue-and-yellow stucco building, figuring a facility with a temperamental toilet, old furniture and broken air-conditioning might have once been a state-of-the-art facility for one of UCLA's most productive and well-known researchers. Slamon explained that he had moved in just six years ago. Before the labs were relocated to the new building, which also houses a few dental offices, a hair restoration clinic and several acupuncturists, they had been scattered around the main UCLA campus just west of Slamon's current location. "Wherever space was available," Slamon explained. UCLA is a public institution, albeit a world-class one, and doesn't have a lot of extra cash. (Operating revenue for UCLA's entire medical center was about half that of Sloan Kettering's in 2017, even though UCLA admitted nearly twice as many patients.)

Thirty years after Slamon published his first article on the *HER2* gene, he said he was still at it, studying why breast cancer patients whose tumors overexpress the protein don't respond to Herceptin and why some HER2-positive patients relapse despite getting the drug. "That's consuming a lot of time," he said. Slamon is also looking for other oncogenes that, like *HER2*, can be treated with targeted drugs, "taking lessons learned from the HER2-Herceptin story to say this can't be the only story like this." Indeed, Slamon's perseverance and creative thinking have found success outside the world of HER2.

In the 1970s and 1980s, British researchers discovered that certain enzymes played a critical role in the step-by-step process of cell division. In the 1990s, scientists at Parke-Davis, a pharmaceutical research division of the drug company Warner-Lambert, had developed a new compound that inhibited the activity of some of these enzymes early in the cell-division cycle, which researchers hoped might slow the growth of tumors. But just as Genentech was initially hesitant to study the drug that became Herceptin, the enzyme-inhibiting compound at Parke Davis also fell victim to internal company politics and larger forces in the world of drug development. Pharmaceutical giant Pfizer acquired Warner-Lambert in 2000, eager to own its cholesterol drug Lipitor. The cancer-related compound, which inhibits the enzyme CDK4/6, fell by the wayside. Delivering the drug, which could be highly toxic, to cancer cells was complicated, and it was not known which types of cancer might respond to it. Plus, Pfizer acquired another, smaller drug company already developing its own cancer drugs. Its CDK4/6 inhibitor sat on the shelf for years, an interesting drug without a use.

But outside researchers, including Slamon, kept experimenting with it, convinced that the drug had untapped potential. Slamon thought perhaps Pfizer's CDK4/6 inhibitor might work against triple-negative breast cancer. Unlike progesterone-receptor-positive, estrogen-receptor-positive or HER2-positive breast cancer, triple-negative had no known driver that could be targeted with drugs. But when Slamon tested the CDK4/6 inhibitor on breast cancer cell lines in his lab, he was surprised to find that cells that were estrogen-receptor-positive and HER2-negative were the ones to respond. Maybe Pfizer's shelved drug had a use after all.

Slamon and a colleague met with Pfizer in 2007 and persuaded the company to launch a small trial at UCLA with a dozen women, all with estrogen-receptor-positive, HER2-negative metastatic breast cancer. The women received the CDK4/6 inhibitor along with a standard antiestrogen aromatase inhibitor. Three of the twelve women experienced significant reduction in their tumors. This got Pfizer's attention. On the basis of the work at UCLA, the company launched larger trials in 2009 and 2013. Federal regulators at the Food and Drug Administration were so impressed with the drug's possibilities that the agency fast-tracked Pfizer's application for approval. The trials proved that the CDK4/6 inhibitor, which Pfizer named Ibrance, slowed or halted tumor growth in more than 80 percent of patients with estrogen-receptor-positive metastatic breast cancer. Best of all, Ibrance was far less toxic than standard chemotherapy. It did not cure women with metastatic breast cancer, but the median amount of time before their disease inevitably began progressing again went from ten months to twenty months. The drug was approved for use in 2015 and has been joined in the marketplace by different CDK4/6 inhibitors made by other pharmaceutical companies. Together, the drugs are the most significant innovation for women with estrogen-receptor-positive breast cancer since hormone therapy. Not surprisingly, research is now under way to study whether the compounds might also be useful for women with early-stage estrogen-receptor-positive breast cancer.

I told Slamon that many people I had talked to had criticized research in breast cancer, which has benefited from more funding than any other type of cancer and yet remains one of the top killers. Despite hormone therapy, chemotherapy, Herceptin and CDK4/6 inhibitors, I said, we seemed far from solving the disease's remaining mysteries, and many patients and advocates I had spoken to were frustrated that progress in fighting breast cancer, while consistent, had been slow and incremental. It took Ibrance some twenty years to come to market, nearly the same time that had passed between Slamon's 1987 HER2 *Science* paper and FDA approval for early-stage patients like me. Slamon told me that breast cancer's complexity and variations mean that, as maddening as it is, incrementalism is the only rational approach. Defeating breast cancer requires attacking it from different angles simultaneously with different tools and techniques, each of which must be invented and refined individually. Along the way, there are profit-driven drug companies, failed lab experiments and ideas that don't pan out. Slamon said he knows that critics of the process mean well. "They are very frustrated with the fact that they put this much money in and there's only a couple of things they could point to that have really changed therapy in breast cancer," he said. "But the flip side of the coin is those things have been big things."

9

from scalpels to sentinels

Most of California was in the midst of a historic drought the spring that I finished chemotherapy. Some nine hundred thousand acres of the state would be charred by wildfires by the year's end. The ground in Los Angeles was cracked and dusty, the hills covered in thickets of long-dead bushes and matted brown grass. In the throes of chemo-induced chemical menopause, I felt drier than the land. Except for my head, I was hairless. The tiny hole in my left arm where my chemo catheter had been remained bloodless but curiously unscabbed.

It was three weeks until my surgery, and Collin and I decided I should rest and recharge far from home. I found a small mountainside meditation center and hotel in Maui where I could do yoga every morning. As we made our way to the airport gate in LA, Evie weaved through groups of travelers, pulling a shiny pink Hello Kitty suitcase behind her, blissfully unaware of what awaited us when the trip was over.

Our room at the meditation center had a four-poster bed made of Indonesian wood and covered with sheets so soft they felt like butter. The first night there, we slept with the windows open and felt the warm Pacific breeze. The next day, I did three hours of hot yoga. I was too sore for another class the following morning, so after breakfast I walked in circles around the property's enormous lime-green lawn. There was a Buddha statue at one end and the ground was moist, the volcano slopes of the island hidden in the mist of heavy clouds. As I paced, I noticed that the dewy morning had attracted some giant snails that appeared to be fixed to the concrete path around the lawn. The snails had enormous swirling brown-and-cream shells. I left to get a cup of coffee from the retreat center's communal kitchen and when I returned, the first snail I had spotted was closer to the path's edge. It had made progress, but so slowly its movement was hardly noticeable.

One night at dinner, a waitress brought Evie some chicken fingers and she pushed the plate away and declared that she wanted a sundae, now! In her toddler disgust, she knocked over a water glass, soaking Collin's lap. He is an eminently patient father, but he scolded her in a stern voice. Stunned, Evie scrunched her eyes for what seemed like minutes of silence before she emitted a wail, tears streaming. I reminded Collin that Evie was tired. I told her to apologize. I put one hand on his leg and held her hand with my other. And then I started to cry too, overwhelmed with the thought that if I was gone and this scene played out again, there would be no one to make peace.

A couple of weeks earlier, after I finished my sixth and final chemotherapy infusion, a nurse had braced herself against my biceps and pulled out the catheter that ran all the way to my heart. Free of it, I could now swim in the ocean. One day in

Hawaii, we drove to a small cove with a black-sand beach where Collin and I had camped years earlier, before we were married. We skipped rocks and Evie dug her small bare feet into the volcanic powder.

The final night of the trip was my thirty-sixth birthday and we celebrated by going to a luau. Evie's eyes stayed fixed on the fire dancers and she beamed when kids in the audience were invited onstage for an impromptu hula lesson. Later, back in our suite, Evie fell asleep on a foldout sofa in the living room attached to our bedroom. Collin and I drank two bottles of champagne and fumbled our way through a night of drunken sex. It was the last time we would be naked together with my body still intact. When Collin and Evie dozed off on the plane the next day, I stared out the window, unable to sleep. My surgery was just four days away.

The English writer Fanny Burney was living in Paris when she began to feel a strange pain in her breast. It was the summer of 1810, and Burney, who had moved to Paris with her French nobleman husband, tried to ignore the ache and the hard spot in her breast that accompanied it. But her husband and friends insisted she see a doctor. The physician who examined her was the most famous surgeon in France and obstetrician to Napoleon's first wife, Empress Joséphine. Just the sight of such an esteemed doctor scared Burney. "I began to perceive my real danger," she later wrote in a letter to her sister. The surgeon prescribed some medicine, but it seemed to make little difference. And so Burney was seen again by the royal surgeon and a team of other doctors. They examined the writer in a room of her house and then asked her to leave so they could discuss her case in private. Burney waited for half an hour before one of the doctors summoned her and delivered the news. The diagnosis was breast cancer. The only remedy with any hope, Burney's doctors told her, was surgical removal of her breast. It would be another forty years before European doctors began using anesthesia in surgery, which meant Burney would have to endure the operation awake and aware.

The doctors prepared her as best they could. In a letter to her sister Esther, Burney described how one member of her medical team had said, "I do not want to deceive you—you will suffer—you will suffer *very much!*" Another asked if she had cried out when she gave birth to her son, Alex. She had. You must do the same during the surgery, he advised. It will help.

Burney had published three novels and given birth to her first and only child at the age of forty-two. She was a strong woman. But there was a good chance she would die of the procedure itself, from a hemorrhage or infection, so Burney got busy writing a will. There was little else to do. Her surgeons, not wanting to frighten her further, refused to tell her exactly when the surgery would take place. Her husband filled a closet with surgical supplies but kept the stash secret from his wife. Weeks passed.

Finally, on the morning of September 30, 1811, a letter arrived informing Burney that the day had come. In a daze, she wandered into the room that had been prepared for the operation, where bandages and compresses were stacked high. Burney began to pace. "I walked backwards and forwards till I quieted all emotion, and became, by degrees, nearly stupid—torpid, without sentiment or consciousness," Burney later wrote to her sister.

A surgeon arrived and gave her a wine cordial. Then the medical team assembled—seven men, all dressed in black. The lead surgeon asked that two old mattresses and an old sheet be brought into the room. When those were in place, he asked Burney to lie down. She began to shake, eyeing the door and windows, thinking of escape. Burney eventually positioned herself on the mattress and a surgeon placed a handkerchief over her face. Through the thin cloth, Burney could see the men gathered around her and the shine of a blade before she squeezed her eyes shut.

When the first cut was made, Burney let out a scream, as she had been instructed. As she would later report to her sister, "I almost marvel that it rings not in my ears still! So excruciating was the agony." Burney could feel the surgeons separate her breast from her chest wall and then meticulously cut free small bits of tissue that remained. The blade scraped against her breastbone and she could feel that too. After twenty minutes, the surgery was over and Burney was carried to bed. Her hands and arms hung from her body, which had gone limp. At least one surgeon's face was streaked with blood.

Fanny Burney was not the first woman to have a mastectomy without anesthesia. There is evidence that women underwent the surgery for hundreds, or even thousands, of years before her operation. But Burney's first-person account of her mastectomy, written in a long letter to her sister, remains the most complete record of such an operation and makes clear what women were willing to endure to banish breast cancer from their bodies. According to the measures of the day, Burney's surgery was a success. She lived another twenty-eight years, dying at the age of eighty-seven. It would be several decades before the advent of modern pathology, so there is no way to know whether Burney's cancer was successfully removed or if the diagnosis itself had been incorrect.

The year that Fanny Burney noticed the pain in her breast, Abigail Adams Smith noticed a small dimple in hers. Smith, the daughter of former president John Adams, was forty-three and lived on a small farm in western New York with her husband and four children. The breast dimple worried Smith, but she hoped it would go away. Instead, it slowly grew into a lump and eventually a large mass. Her breast was distended and misshapen by the time she wrote to her parents in 1811 and told them a local doctor had diagnosed her with breast cancer.

Smith's parents urged her to join them in Massachusetts so she could get medical advice in Boston. Once at her parents' home, she wrote to Benjamin Rush, the country's most respected doctor and a signatory to the Declaration of Independence. Smith described the appearance and feel of her breast. What should she do? Rush never wrote her back, but he did send a letter to her parents, asking them to break the news to their daughter gently. "Let there be no delay in flying to the knife," Rush wrote. "It will be but a few minutes if she submits to have it extirpated... It shocks me to think of the consequences of procrastination." The operation was scheduled for October 8, 1811, a week after Fanny Burney's breast was removed in Paris, more than three thousand miles away.

On October 7, surgeon John Warren examined Smith and explained what would happen the next day. Smith would sit in a reclining chair and extend her arm above her head. Warren or his assistants would tie her down by her waist, legs, feet and other arm. When she was ready, Warren told her, he would pierce and lift her breast with a fork-like instrument and slice it from her body with a razor. Immediately after, he would remove a red-hot spatula from an oven and apply it directly to the wound, cauterizing it. Finally, he would stitch her up and apply bandages. The whole thing would take about half an hour.

The next day, Smith climbed the stairs of her parents' house in Quincy, Massachusetts, and entered the bedroom where Warren and his assistants were waiting. She would not need to fully undress for the operation, just slip her arm out of her clothing, exposing her cancerous breast. Smith was dressed in her Sunday best, which would be stained with blood and soaked with sweat by the time the surgery was over. In a letter to a friend, Smith described the surgery as a "furnace of affliction." She survived the procedure but died of metastatic breast cancer two years later.

Unlike cancers that develop in the body's internal cavities, malignancies of the breast are easy to access, which made them prime targets for surgery in the early nineteenth century. Back then, physicians hoped that mastectomies, like the ones Fanny Burney and Abigail Adams Smith endured, would remove the threat of breast cancer for good. But even the most optimistic doctors knew it was a long shot. Often, the disease spread anyway and became fatal. Exactly how it spread was a mystery. In 1889, an English doctor named Stephen Paget published a paper in the *Lancet* theorizing that breast cancer cells spread through the bloodstream and set up metastatic outposts in specific distant organs especially hospitable to cancer growth. William Stewart Halsted, a young American surgeon just starting his career when Paget's paper was published, put his faith in a different theory. In doing so, he sent the trajectory of breast cancer treatment in the wrong direction for more than half a century.

Halsted was born in 1852, the eldest son of a wealthy New York City businessman, As a young boy, he was educated at home by private tutors and later attended boarding school, then Yale, then, finally, the Columbia University College of Physicians and Surgeons. All medical students must learn anatomy, but for Halsted this part of his studies became an obsession. He bought extra cadavers with his own money and dissected them after class. He even operated on dogs to practice his surgical techniques. There was always more to learn. The field of surgery was advancing with a velocity no one in medicine had seen before. Anesthesia was in use, along with operating-room antiseptic techniques to reduce infection, and American surgeons were attempting procedures they would never have considered in Fanny Burney's day.

The most innovative surgeons, though, were in Europe. Halsted traveled to the Continent in 1878 to study with some of the world's best and returned home to New York in 1880, the same year the American Surgical Association was founded. Halsted was soon hired to operate at several hospitals in New York City. With his training at Columbia and in Europe, Halsted quickly gained a reputation as a highly capable surgeon and began teaching medical students in the evenings when he was done working at hospitals. He raised money to build a then state-of-the-art surgical facility at Bellevue Hospital, where he operated and oversaw other doctors using many of the techniques he had mastered in Europe.

In 1884, Halsted read an article in a medical newspaper

about a plant derivative that could produce a numbing effect when injected into nerves. Excited by this innovation and its surgical applications, Halsted wanted to get his hands on the derivative as soon as possible. The substance was cocaine. At the time, it was legal and used in a variety of elixirs and drinks sold in pharmacies. Halsted ordered some pure cocaine from a pharmaceutical company and began experimenting with the drug, injecting his medical students with it to test the effects. Within a month of acquiring the drug, Halsted was using it to numb patients for minor surgical procedures and extolling the virtues of cocaine to other members of the New York medical community. Word spread rapidly. The drug became especially popular among New York dentists, who used it to numb the mouths of patients before tooth extractions. In addition to experimenting with cocaine on students and patients, Halsted also experimented on himself. Soon, colleagues began noticing changes in the doctor. They saw that Halsted's hands sometimes shook; he spoke more quickly and sometimes didn't make sense. In his zeal to incorporate a new anesthetic into modern surgery, Halsted had become a cocaine addict.

In 1886, in the throes of his drug dependence, Halsted took a leave from hospital duties. He went back to Europe, hoping to reset his system, and even embarked on a weeks-long sailing trip he thought might function as a detox. On the journey, Halsted reportedly broke into a medical cabinet on board and retrieved some cocaine. Halsted went to rehab twice at a hospital in Rhode Island but emerged from the experience with a new problem. To counteract the jittery effects of cocaine, he had begun using morphine, a central nervous system depressant. Halsted would remain an active drug user for the rest of his life, mysteriously disappearing from work for months at a time and worrying surgeons who operated alongside him.

Miraculously, Halsted's addictions did not slow his career trajectory. On the strength of his medical reputation in New York City, Halsted was appointed head surgeon at the newly opened Johns Hopkins Hospital in Baltimore, just two years after his second stint in rehab. That he could feed his addiction and handle a prestigious job with immense responsibility is a testament to Halsted's skill as a surgeon. At Johns Hopkins, which would grow into one of the country's premier hospitals, he was a surgical rock star, pioneering scores of new techniques, from orthopedic operations to procedures on the stomach, gallbladder and thyroid. He introduced operating-room rules that, at the time, were revolutionary, requiring surgeons to wear clean white outer garments instead of the formal suits that were customary and introducing rubber gloves to the surgical suite. Halsted had been known in New York City medical circles as a surgeon who operated quickly and decisively, but at Johns Hopkins he earned a reputation as a methodical and meticulous operator. Known as "the Professor," he spent hours on surgeries, sometimes finishing procedures with hundreds of clamps attached to blood vessels that had to be sewn closed one silk suture at a time.

When it came to breast cancer, Halsted's techniques were guided by his belief in the prevailing centrifugal theory of the disease—that it started in a woman's breast and spread directly outward into the body. Paget's systemic theory of cancer was still an idea on the margins. Chemotherapy was not even a notion. There was a single treatment for breast cancer patients who made their way to Halsted in Baltimore—surgery, and the more extensive, the better.

While women in Halsted's day were mercifully anesthetized during surgery, they often woke up in bodies they did not recognize. Halsted promoted a type of breast surgery far more extensive than the kind performed on Fanny Burney and Abigail Adams Smith. Halsted's operation, which was also performed by some doctors in Europe, was known as the radical mastectomy. *Radical* comes from the Latin word for "root," and the theory behind the radical mastectomy was that, by removing the tumor and a wide area around it, a surgeon could eliminate all cancer cells in the body and, therefore, the risk of recurrence.

In his radical mastectomies, Halsted removed not just the breast but also skin, muscles beneath the breast, large clumps of lymph nodes in multiple areas of the chest and even some bone. Halsted did his radical mastectomies "en bloc"; that is, keeping all tissue intact and connected. He worried that cutting through cancerous tissue could cause cancer cells to break free and reseed themselves in nearby tissue. The surgeries, which took hours, often left women with concave voids in their chests. Many lost the use of one arm. But breast cancer was so feared and so consistently lethal that patients flocked to Halsted and Johns Hopkins. Word of his purported success got around. Halsted claimed that fewer than 10 percent of his patients saw their breast cancer recur in the area of the operations. Such low local recurrence rates were astonishing, a fraction of the rates reported by other surgeons in the United States and in Europe.

Years later, it became clear that Halsted's figures were off.

His scalpel saved the lives of some breast cancer patients, but many others later died of the disease. Halsted followed women for just a few years, even though cancers can return much later and elsewhere in the body. But by ably and, in many cases, safely excising cancerous breasts and lymph nodes, Halsted achieved something remarkable. He had managed to, if not save, then certainly extend the lives of some of his patients. Medical historian Barron Lerner wrote that, because of Halsted, "breast cancer evolved from a disease that recurred visibly and rapidly into one that returned invisibly and more slowly."

Halsted died in 1922. By the early 1930s it was clear that the success of his radical mastectomies had been overstated, but Halsted's influence did not wane. Instead, it reverberated in operating rooms across America. When breast cancer recurred after a radical mastectomy, many surgeons believed the operation had simply not gone far enough. Doctors who operated in the hospital now known as Memorial Sloan Kettering took Halsted's theories and methods to the limit; in the 1950s, surgeons performed "extended radical mastectomies," in which they removed multiple sets of lymph nodes, and "super-radical mastectomies," in which they excised, in addition to the breast, muscles and lymph nodes, parts of the sternum, clavicle and chest wall. The surgeries were horribly disfiguring and sometimes fatal, but to some American surgeons the risk was worth the supposed reward. To them, Halsted was a hero, not a butcher.

But not every surgeon was a Halsted devotee. In the 1920s, a British doctor named Geoffrey Keynes, the brother of famed economist John Maynard Keynes, bucked the Halstedian worldview to experiment with smaller and shorter breast cancer surgeries that he combined with implanted radium, a relatively new treatment option at the time. In 1896, Emil Grubbe, working in Chicago, had become the first physician to treat a breast cancer patient with radiation. He reported that the treatment shrank the woman's tumor, although she later died of metastatic disease. In the years after Grubbe's work, radiation gradually became a complementary treatment to breast cancer surgery. The technology, in which radiation is directed at the tumor, works by damaging cell DNA. Cancer cells, which divide more often and more rapidly than other cells in the body, are particularly vulnerable to radiation therapy.

Keynes's approach appeared to save lives at a rate similar to Halsted's, but for decades few in the surgical community paid it much attention. Then, in 1953, a Cleveland surgeon named George Crile Jr. heard a lecture on Keynes's work. Crile had long harbored doubts about the wisdom of the Halsted mastectomy and the super-radical mastectomies some influential New York City surgeons favored. (Crile's father, also a surgeon, had found that patients subjected to radical surgery often went into shock.) Crile preferred a surgery then known as the partial mastectomy, a procedure in which the cancerous breast tumor was removed but chest muscles and bone were left alone. Halsted followers regarded Crile's surgeries as dangerous, but there was no comprehensive data to prove which approach was best. A long-term clinical trial was the only way to know. Meanwhile, tens of thousands of American women continued to undergo radical mastectomies every year. Some were cured; many were not.

The first modern American research effort to test the efficacy of the radical mastectomy began in 1971, nearly ninety years after Halsted performed the procedure for the first time. The federally funded National Surgical Adjuvant Breast and Bowel Project (NSABP)—the same research effort that had helped prove the value of chemotherapy and, later, the safety of eliminating that treatment for some women—launched a trial on breast cancer surgery. Women whose breast cancer had spread to their lymph nodes were randomly selected to undergo either a radical mastectomy or a simple mastectomy, which removed only the breast. Women in the simple-mastectomy group also received radiation therapy. Women whose lymph nodes were free of cancer received one of three treatments: a radical mastectomy, a simple mastectomy with radiation, or a simple mastectomy with no radiation.

Recruiting women to join the trial was not easy. The radical mastectomy was part of standard breast cancer care and many women did not wish to gamble with an unproven treatment, even if it meant they could keep more of their bodies intact. Aggressive surgeons, like some at Sloan Kettering, said the trial was a folly that risked women's lives. They refused to participate and steered their patients away. Crile, though, supported the trial and gave television interviews about the debate, even publishing a book in 1973 called *What Women Should Know About the Breast Cancer Controversy*. Crile was so committed to the idea that smaller surgeries were no less effective than larger ones that he maintained his position even after his first wife was diagnosed with breast cancer, had a nonradical mastectomy and later died of metastases. Second-wave feminists who had led the charge to legalize abortion nationwide in 1973 voiced their support for the mastectomy trial, and women's magazines published articles urging women to question the standard procedures and become more involved with their own health-care decisions. (In 1974, Crile's second wife also developed breast cancer and was successfully treated with a lumpectomy, in which only her tumor and a small bit of tissue around it were removed.)

Bernard Fisher, a strong-willed and charismatic Pittsburgh surgeon, chaired the NSABP. Fisher had performed many radical mastectomies, but over the years he had begun to doubt the procedure's effectiveness and the theory underpinning it, that breast cancer spread in a centrifugal pattern outward from an initial tumor. Fisher, like Stephen Paget in the 1880s, believed it was more likely that breast cancer cells broke away from primary tumors and spread through the bloodstream, not in an orderly and predictable pattern that could be defeated with a scalpel. By this logic, a smaller surgery that removed only a cancerous breast tumor, leaving muscle and other tissue alone, might be just as effective as a radical one.

As the head of the first clinical trial to test this theory, Fisher drew the ire of his fellow surgeons. They said he was dangerous and castigated him in public and at scientific meetings. But Fisher held fast. A randomized clinical trial removed surgeons from the decision-making process, but it was the only way to produce clean data that could be trusted. By 1981, results from the trial showed that women who had radical mastectomies fared no better than those who had simple mastectomies. The Halsted era of breast cancer surgery might have ended right there, but old habits die hard. Some surgeons, particularly those in Halsted's hometown of New York City, continued to routinely perform radical mastectomies for another decade.

A second, bolder NSABP trial went even further than the first, demonstrating in 1985 that, in many cases, there was no need for even a simple mastectomy. This second trial showed that lumpectomies along with radiation treatment proved just as effective as simple mastectomies when it came to preventing recurrence. Instead of a woman undergoing a biopsy immediately followed by a mastectomy in a single procedure, in the 1970s and 1980s women increasingly had the opportunity to hear the results of their biopsies and then set the terms of their own surgeries.

Mastectomies still had a place in breast cancer treatment. Sometimes a tumor was too large to remove via a lumpectomy. In addition, doctors continued to surgically remove large sections of lymph nodes in women's armpits. The lymphatic system, part of the immune system, carries a fluid called lymph throughout the body. Sometimes, cancer cells can enter the lymph vessels and spread to lymph nodes the same way they spread to distant organs via the bloodstream. But on the heels of Fisher's work, surgeons began to wonder whether so many lymph nodes truly needed to be excised. Removal of large numbers of lymph nodes can cause complications, including chronic and painful arm swelling. In the 1990s, a Los Angeles breast cancer surgeon named Armando Giuliano mapped the lymphatic systems around breasts and determined that the lymph nodes most likely to contain cancer could be pinpointed. Known as "sentinel nodes" and first identified in the 1970s, these small nodules can be removed during surgery and examined under a microscope immediately for signs of cancer. Giuliano theorized that cancer-free sentinel lymph nodes indicated that the other lymph nodes around a woman's breast were also cancer-free and could be left alone. Like Fisher, Giuliano faced criticism from his fellow surgeons, who were suspicious of the sentinel-node biopsy and felt it was unnecessarily risky. Again, randomized clinical trials provided clarity. Studies showed that Giuliano was correct. Today, when surgeons suspect a woman's lymph nodes might contain cancer, she typically undergoes a sentinel-node biopsy. Sometimes, cancerous sentinel nodes lead surgeons to remove a large number of other nodes. But increasingly, doctors are forgoing this step in favor of radiation treatment, which can be just as effective.

Monica Morrow, the chief breast surgeon at Sloan Kettering, became a doctor right around the time that Bernard Fisher's NSABP trials upended the norms of traditional breast cancer treatment. "When I started my surgical residency, people were still arguing about whether it was safe to do a modified radical mastectomy [in which a breast and lymph nodes were removed but not muscle] instead of a radical mastectomy," Morrow told me when we met in the summer of 2017 at the hospital's breast center. Morrow remembers that early in her career, it was breast cancer patients themselves who pushed for changes in the operating room, lobbying the government to study the issue and encouraging other women to stand up for the integrity of their bodies and advocate for smaller surgeries. "Today, you have the complete converse of that," said Morrow.

While breast cancer chemotherapy rates have fallen and evidence continues to pile up proving the safety of smaller surgeries, women today are pushing their surgeons in the opposite direction, asking for—and, in some cases, demanding—more treatment instead of less. Decades of research proving that large surgeries are often unnecessary can fall by the wayside in the exam room.

For example, in recent years the number of women who qualify for lumpectomies but undergo mastectomies has been increasing. A study coauthored by Morrow and published in the *Journal of Clinical Oncology* in 2005 looked at surgical decision-making among women with DCIS and small invasive cancers. These patients, based on scientific data, could safely opt for either lumpectomies with radiation or mastectomies. In cases where a surgeon was the primary decision-maker, the rate of mastectomies among white women (who represented 70 percent of the sample) was 5 percent. When these women and their surgeons shared equally in the decision, the rate was more than three times higher. When women were the primary decision-makers, the rate was more than five times higher than when doctors drove the choice. "So it's clearly a patient-driven thing," Morrow said. Another study conducted by Morrow and collaborators showed that about half of women with DCIS and stage I invasive breast cancer do not realize that they have the same chance of surviving whether they have a mastectomy or a lumpectomy with radiation. "Which is, of course, the one thing you need to know to really make an informed decision," said Morrow.

Some women may choose to have mastectomies because they are afraid to have the radiation treatment that often accompanies lumpectomies. But here, too, perceptions of risk do not mirror reality. In a study of UCLA breast cancer patients treated with radiation after surgery, published in the journal *Cancer* in 2018, researchers wrote that 85 percent said the experience was "less scary" than they'd anticipated. A majority of those who had lumpectomies said short- and long-term side effects of radiation treatment (including pain, fatigue and changes in appearance) were less severe than they had expected.

Alongside misunderstandings, there are legitimate nuances that can complicate the decision of whether to have a mastectomy or a lumpectomy. Breast cancer recurs in fewer than 10 percent of women who have had lumpectomies, but when it happens, these cancers are typically treated successfully because women who've undergone breast cancer treatment are monitored closely for recurrences or new cancers. Overall survival is the same, but a second bout of the disease could mean further surgery, radiation or chemotherapy. Many women do not want to face that possibility, however remote it may be. Just the stress of being monitored through annual or more frequent mammograms is enough to lead some women to decide to have an entire breast removed. They may have a strong desire to put their breast cancer experience behind them, even if it means their bodies will carry on in a more altered state. These are not scientific decisions but emotional ones.

Likewise, women diagnosed with cancer in one breast and at very low risk of developing disease in the other breast are also now more likely to opt for double mastectomies than they were in years past. A 2017 study published in the *Annals of Surgery* found that rates of such double-mastectomy surgery nationwide tripled between 2002 and 2012.

In my case, lumpectomy was not an option. In addition to the two invasive tumors in my left breast, I had widespread DCIS. By the time a surgeon cut out all of my disease, there would have been little left of my breast anyway. I understood that the chances of developing a new cancer in my other breast were very low, although that risk would compound every year and I was only thirty-five. Over the course of a normal life span, my doctors told me, I had about a 20 percent chance of developing breast cancer on the other side. That was more risk than I was willing to accept, so I opted to have a double mastectomy. I was a fully informed patient who nonetheless chose a larger surgery than what was medically required.

After I opted for a double mastectomy, there was the matter of my lymph nodes to consider. At my diagnosis, biopsy results showed that at least two nodes in my armpit had been invaded by cancer. I technically qualified for a sentinel-node biopsy and would undergo radiation treatment in the area anyway, which can kill off remaining cancer cells. Plus, I had already had chemotherapy and HER2-drug treatment. But several surgeons I consulted still recommended I undergo a "full axillary dissection," meaning removal of the primary clump of lymph nodes in my armpit. They cited my young age and the fact that I had a fast-growing cancer, saying the risk of leaving some nodes behind made them uncomfortable. I had read the medical literature and knew it was a close call, but I decided to have the dissection. I was driven not by evidence but by what decision-science researchers call "anticipated regret." Quite simply, I did not want to shoulder the blame if I ended up having a recurrence or a new cancer. The most common complication of a full axillary dissection is a condition called lymphedema. Because lymph nodes are removed, lymph does not flow as well in the arm, so it can swell and be painful. Lymphedema is more common in women who are overweight, and I was not, which helped make the decision easier.

I understand why even fully informed breast cancer patients

sometimes choose larger surgeries when small ones would suffice. It's not logical, but it's the way we are wired, literally. "We are animals and our reactions are more powerful than our cognition," said Steven Katz, a medical-decision scientist at the University of Michigan. "We have very powerful intuitive and emotional reactions that have saved the species through millennia. Those reactions are very powerful in the exam room." Katz, a trained internist who studies breast cancer decision-making, frequently collaborates with Monica Morrow to study trends in breast cancer treatment and the factors that drive decision-making. In particular, the two are interested in how to reduce overtreatment in surgery.

In a 2017 study published in *JAMA Oncology,* Morrow, Katz and several other authors studied what happened after new clinical guidelines came out addressing variation in what constitutes a successful lumpectomy. In addition to removing a tumor, a surgeon performing a lumpectomy also removes a small amount of tissue around it. After surgery, a pathologist examines this tissue, known as a margin, under a microscope to make sure that it does not contain cancer cells. For a long time, surgeons disagreed on how wide margins needed to be to ensure a woman's breast cancer was completely removed. Some recommended women have additional surgery if the margin seemed to be too thin. But after new clinical guidelines were published standardizing margin widths, Morrow, Katz and their coauthors found that the rates of mastectomies went down. It turned out that women who had lumpectomies but were later called back for additional surgery after doctors said their margins were too thin often opted for single or double mastectomies. The knowledge that a threat existed for what felt like a second time ratcheted up their primal fear and they reacted strongly. "We were able to show that by changing the guidelines in accordance with modern practice, we could reduce that surgery," said Morrow.

In addition to patient preferences, the other major factor driving women toward larger surgeries is insurance coverage. A 1998 federal law, passed during the golden age of breast cancer activism in the United States, requires insurance companies to cover single or double mastectomies and reconstruction even in cases where clinical guidelines say that a lumpectomy is appropriate.

Still tan from Maui, I sat on a gurney swinging my legs over the side. I had changed into a blue hospital gown and brown socks with no-slip bottoms, then folded my street clothes and set them in a neat pile on a nearby chair. I had read about Abigail Adams Smith and, as I tucked the sides of the hospital gown underneath me to keep out a draft, I thought of how she had worn a nice dress to her mastectomy.

A nurse was starting an IV in my right arm when my surgeons arrived. Dr. Amy Kusske, a no-nonsense surgical oncologist with short feathered hair, would perform the first part of my surgery, amputating my cancerous breast and removing a clump of nearby lymph nodes. UCLA's chief of plastic surgery, Dr. Andrew Da Lio, would then take over, amputating my healthy breast and starting to create a set of replacements. After I signed a surgery consent form, Da Lio asked me to lower my hospital gown. Using a purple marker, he drew dotted lines around my breasts and around my left nipple, the one that had oozed goo months earlier. When he was done, my breasts looked like they were covered in geometric bruises. I pulled my gown back up over my chest and Collin hugged me as my eyes welled up with tears. I clenched my jaw. As the doctors and nurses wheeled the gurney down the hallway and into the operating room, a sedative delivered through my IV reached my brain. The last thing I remember is lying on my back and feeling my arms being strapped to boards, crucifix-style.

I woke up six hours later in a recovery room, bundled in blankets and in excruciating pain. After a dose of narcotics, I was wheeled into a private hospital room, where I drifted in and out of sleep until the next morning when Justin Maxhimer, one of the plastic surgery department's chief residents, arrived to check my incisions. As he folded down my blankets and hospital gown, I noticed for the first time that my chest was covered in gauze that had been intricately wrapped around my torso. Maxhimer used a pair of scissors to cut through the gauze and gently removed the covering. Where my breasts had once been were two small mounds, the artificial breasts Da Lio had begun building the day before and would finish in a second surgery many months from now. The incisions, one across my left breast and another underneath my right breast, were clean, straight and covered by small white strips. "Everything looks good," Maxhimer said, helping me sit up slightly so he could wrap a sports-bra-like post-mastectomy garment that closed in the front with Velcro around my torso. It was tight and meant to compress my chest so fluid could not build up inside it. I could go home in a few days.

About a week later, my doctors said my body was healing nicely, but the pain had not subsided much. It shot through my upper body like a barbed dart every time I shifted in a chair. I hadn't been able to make it into bed yet. At night, I drifted in and out of opiate-induced sleep in a recliner set up in the middle of our living room.

Three surgical drains shaped like hand grenades hung from my torso. The drains were plastic reservoirs connected to clear tubing through which fluid that built up inside my body could run out. Several times a day, I had to unscrew the bulbs from the tubing and empty the fluid into a graduated cylinder to measure the output. The fluid was red the first day and turned progressively pinker and then yellow as my insides healed.

I was itchy and my hair was greasy, but the drains and the incisions meant a shower was out of the question. My father's two sisters, both registered nurses, had flown to Los Angeles from upstate New York to care for me in the days after my surgery. They promised a quick sponge bath would make me feel better.

My aunts helped me slowly remove the mastectomy bra. Exposed to the air for the first time in days, the skin on my chest and back felt cool. I inhaled deeply and felt the barbed dart twist inside me. I hunched over the bathroom sink, the drain bulbs in my right hand, my left hand on the counter to keep myself steady. My aunts gently ran warm soapy washcloths across my back and over my shoulders. I felt vulnerable and deformed, but the soap smelled fresh.

Just as I started to think about getting wrapped up and back in my recliner, I heard my cell phone ring in the other room. Collin answered it and pushed the bathroom door open. "It's Dr. Hurvitz and she said to put the phone on speaker." I turned from the sink, drains in hand, and we all huddled around the glowing iPhone screen. "You had a pathologic complete response,"

Hurvitz said. "There was no cancer found in any of your breast tissue or any of the twenty-two lymph nodes removed. Yay!"

As one of the patients in the Kadcyla trial who had achieved a pathologic complete response, I'd had the best outcome a breast cancer patient could hope for. If the tumors in my breast and lymph nodes were gone, there was a good chance no cancer remained anywhere else.

Collin stepped out of the bathroom to call the rest of my family with the good news. My aunts rewrapped and dressed me. I hobbled back to the recliner and slowly flipped up the footrest. There were no guarantees of cure, but the pCR was as close as I could get. It meant I would most likely survive my breast cancer ordeal. I lay still, waiting for my recent dose of painkillers to kick in, and let the relief wash over me. But a disturbing thought crept into my mind. If the cancer had been killed off before my operation, that meant the procedure had served no real purpose. Had I needed the surgery at all?

This is a question breast cancer surgeons are increasingly asking. Traditionally, breast cancer treatment has been delivered in a specific order: surgery, chemotherapy, then radiation. As more women undergo chemotherapy and other drug treatment before surgery, doctors are learning more about how their patients' cancers respond. Studies have shown that women with triple-negative or HER2-positive breast cancer who achieve pCR, so-called exceptional responders, are more likely to survive than women with these subtypes who do not have pCR. (My odds of survival were boosted to more than 90 percent once I achieved pCR.) Several trials are now under way to study whether exceptional responders might safely avoid surgery. (The correlation between pCR and survival is less significant among women with HER2-negative, estrogen-receptor-positive disease, and the notion that these women might be able to avoid surgery is not currently being studied.)

Dr. Henry Kuerer, a breast cancer surgeon at M. D. Anderson, is running one such trial. Kuerer told me that early in his career, he had a patient with a large malignant breast tumor that had spread to her lymph nodes. She was treated with chemotherapy first in hopes that the tumor size could be reduced before surgery. After Kuerer removed her breast and lymph nodes, a pathology report showed that she had a pCR. Research had recently shown that women who achieved pCR were more likely to survive, and Kuerer was thrilled to share the news with his patient. But his patient was not thrilled. She wanted to know why Kuerer had removed her breast if it was cancer-free. "I have spent the last thirty years trying to work this out," Kuerer told me.

In the 1980s and 1990s, several surgeons (though not Kuerer) experimented with skipping surgery for patients whose cancers appeared to respond well to chemotherapy, offering radiation treatment in place of scalpels. The studies did not go well. "They didn't have good imaging and they never did needle biopsies and there were high local recurrence rates," Kuerer said. Kuerer and his colleagues studying whether surgery can be eliminated for exceptional responders are mindful of this history and are proceeding with a great deal of caution.

Surgeons were loath to give up the radical mastectomy. The bar for giving up all surgery, even for a small, carefully selected share of breast cancer patients, is likely to be even higher. In breast cancer, patients and doctors are far more willing to accept the addition of treatments that might improve outcomes than the removal of treatments that do not. "You go to a meeting and they say, 'Oh, there's a two percent difference in disease-free survival by adding this new drug,' and everybody goes home on Monday and starts prescribing it. More is always better in the American way of thinking," Morrow told me. "Doing less always takes a long time."

The only way to determine if some breast cancer patients can safely skip surgery is to have them do just that and observe what happens. This is risky, so before such a trial could begin, Kuerer and his colleagues had to prove that doctors could accurately predict pCR without doing surgery. In a study they published in the *Annals of Surgery* in 2018, forty women with triple-negative or HER2-positive breast cancer who had already received chemotherapy and/or other systemic drug treatment underwent biopsies just prior to surgery in which an average of twelve tissue samples were taken. (Kuerer told me he was surprised so many women agreed to participate in the trial, given that they would see no benefit personally.) Following the surgery, pathology studies of the removed tissue were compared to the pCR findings of the pre-surgical biopsies. After a statistical analysis, Kuerer and his colleagues found that they could predict pCR with 98 percent accuracy. On the strength of this study, Kuerer launched a trial at M. D. Anderson that will eliminate surgery for fifty women with HER2-positive or triple-negative breast cancer identified as having pCR after chemotherapy (plus Herceptin for women with HER2-positive disease). Women will still undergo standard radiation treatment.

"We need to be exquisitely safe," Kuerer told me. Patients are deemed ineligible for the study if imaging or biopsy results indicate they have residual invasive cancer or DCIS. Trial results will likely be available by 2021. At that point, Kuerer will know how many women in the trial who did not have surgery remained cancer-free. Even if the trial produces a perfect result, with all fifty women alive and cancer-free after several years, Kuerer knows it will not be enough to change standard practices nationwide. Far more studies with far more women and good results will be needed for that. Kuerer said even if further research shows that surgery can be safely eliminated for some patients, he expects resistance from doctors akin to the kind Bernard Fisher faced when his study demonstrated the safety of lumpectomy.

"Some doctors will say, 'Why not leave good enough alone?' There are financial implications of this for someone whose livelihood depends on surgery," Kuerer told me. He said the leader of a surgery trade organization heard about his pCR trial and called Kuerer to inquire how the study results might affect surgeons. "We, as physicians, really need to take a higher ground on this. Instead of saying, 'What's in it for us?' we need to say, 'What's in it for the patient?'"

10

whole again

Recovering from my double mastectomy and initial reconstructive surgery, I felt, for the first time, like a very sick person. Work was out of the question. I could barely walk in the days after my surgery and, for more than a month, could not drive. When I rode in the car, wearing a seat belt was so agonizing that someone had to tuck the shoulder strap behind my headrest, away from my body. My pectoral muscles, full of stitches, throbbed despite the opiates and muscle-relaxing drugs I took around the clock. I love to cook and tried to make a meal one evening a few weeks after my surgery, hoping the process might distract me from my pain. But I found that pulling out drawers in the kitchen hurt. Cutting vegetables hurt. I moved my arms and shoulders as little as possible, which left me so stiff I wondered if I would ever regain a full range of motion. Worst of all, I could not pick up three-year-old Evie.

About six weeks after my surgery, I went to a Dodgers game with my family. It was a day game and we sat in the stands under the hot sun for hours. As we stood to leave at the end of the ninth inning, I noticed that the back of my left hand was puffy, the tendons running to my fingers invisible. My left wrist was also wider than my right. I had developed lymphedema, a serious condition made worse by the midday heat. The twenty-two lymph nodes removed during my surgery had left the liquid lymph with no place to go, and it was pooling inside my arm and hand. The radiation treatment I received after surgery had exacerbated the problem. I ordered custom-made compression gloves and sleeves that would keep the swelling down, and I found a physical therapy practice in Beverly Hills that specialized in post-mastectomy patients. Several times a week, therapists massaged my upper body, loosening my muscles and coaxing the lymph inside my arm and hand to drain. The therapists also taught me exercises I could do at home to regain my strength and flexibility.

Over time, I did become stronger and more flexible. For decades, doctors had advised post-mastectomy and reconstruction patients to avoid serious exercise out of concerns about further injury, but recent studies have shown that, once the wounds heal, vigorous exercise can safely improve a patient's strength and quality of life. The lymphedema, which my breast surgeon suspected would be temporary, eventually disappeared. After about six months, I felt strong enough to scoop Evie into my arms.

My chest itself would be forever numb, but at least it looked quite nice. Da Lio, my plastic surgeon, had used silicone implants to construct a new set of breasts for me that, when clothed, appeared entirely normal. My scars faded and my breasts were perky, symmetrical and relatively soft. My reconstructed chest bore no resemblance to most of the breast cancer–surgery images I had seen online in the weeks leading up to my operation.

In the Halsted era, women who underwent radical mastectomies were left with mangled torsos and widespread scarring. Sometimes surgical wounds remained open to heal on their own. In other cases, skin was grafted from other parts of women's bodies to cover the area where breasts had once been. There was little thought given to making breast cancer patients physically whole again. Even after the invention of silicone breast implants in the 1960s, women who had surgery for breast cancer were often left flat. They stuffed socks in their bras to hide their amputations or were issued breast prosthetics. Today, more than one hundred thousand reconstructive surgeries are performed on U.S. breast cancer patients every year. Nearly 40 percent of women who have mastectomies opt to have plastic surgery and can choose from an ever-growing menu of implants and techniques.

As the beneficiary of modern breast reconstruction, I had a lot to be grateful for. But the reporter in me was disappointed. Although I had researched the advanced reconstruction techniques used in my surgery, I had been not been able to document the process myself.

And so, on a spring day in 2018, I arrived at the Ronald Reagan UCLA Medical Center at noon. I removed my street clothes, carefully folded them and placed them in a pile the same way I had on the day of my mastectomy. But instead of putting on a hospital gown, I changed into a set of bluish-green scrubs. I was at the hospital to observe the surgery of a woman undergoing nearly the same procedure I had had nearly three years earlier. Da Lio, who had arranged my visit, arrived on the

OR floor a few minutes after me, rushing from the hospital cafeteria where he had just gobbled down some soup. This would be Da Lio's second reconstructive surgery of the day. After giving me a quick hug, he led me over to a tall metal rack in the hospital hallway that held boxes of surgical accessories. He tied a paper hair covering neatly around his head and looked as handsome as a television doctor. I put my surgical facemask on upside down, which made Da Lio chuckle as he stepped behind me and tied it properly.

Now in his mid-fifties with salt-and-pepper hair, Da Lio has spent his entire career at UCLA. He performs cosmetic surgeries—face-lifts, tummy tucks, and the like—but as chief of plastic surgery at an academic medical center, more than half of his time is spent operating on breast cancer patients. In a way, he has been training for the work nearly his entire life. Da Lio's mother, Domenica, an Italian immigrant, was a pattern-maker and seamstress. She spent hours every day at her manual sewing machine, pumping a pedal on the floor that operated the contraption's needle. When her ankle grew tired, her young son would pump for her. He was always at her side. She taught him how to sew waistlines and how to cut fabric to fit women's bodies, sketching patterns on paper that looked a lot like the lines Da Lio had drawn on my body before my mastectomy. "I remember my mother teaching me how to make a bra," he told me. Deconstruct a bra and you get the same pattern that plastic surgeons draw on women having breast-reduction or breast-lift surgery.

When he was a teenager, Da Lio saw a television show about Alma Dea Morani, the first female member of the American Society of Plastic and Reconstructive Surgeons. Morani's parents were Italian immigrants, like Da Lio's. "So I knew what I wanted to do," he said. After medical school at the University of Michigan and residencies at Dartmouth and the University of Southern California, Da Lio took a job at UCLA, where advances in surgery allowed doctors to perform breast-reconstruction operations that had once seemed impossible. The chief of the plastic surgery department back then was Dr. William W. Shaw, who had made headlines in 1979 when he operated on a young New York City woman whose hand had been severed in a subway accident. Shaw reattached the hand and restored enough of its function that the woman, a trained musician, was able to play the piano afterward. Shaw's talent for microvascular surgery, in which tiny blood vessels are sewn together to create blood flow in reattached limbs and flesh, allowed him to fashion breasts out of flesh harvested from patients' abdomens, buttocks or backs. "When I was doing microsurgical breast reconstruction back in the nineties," Da Lio said, "we would have bleachers behind us of people looking down at Bill Shaw and me, learning how to do the techniques we were presenting." Known as autologous reconstruction, such procedures are now done at many hospitals across the country and offer patients an alternative to silicone or saline breast implants. (The recovery period for an autologous-reconstruction surgery is longer and more painful than it is for implant reconstruction.)

There was no separate observation area above the operating room where I would witness breast cancer surgery firsthand. I would be in the room itself, which was smaller and more crowded than I expected. It had been scrubbed down and restocked after an earlier procedure. Shiny instruments were spread across several tables, and doctors and nurses scurried around them. Holding my notebook and pen and feeling a bit nervous, I took a deep breath. The air felt cool and clean. By state law, to reduce the presence of airborne microbes, the air in California operating rooms must be circulated out and replaced with new, clean air twenty times an hour. I could feel a slight breeze as the system hummed above us. Bright lights and cameras hung from the ceiling on heavy, adjustable arms that snaked down over the operating table. They were smooth, round and looked like small spaceships. I counted eight screens around the room, including computers and video monitors that would project a close-up view of the surgery onto a wall.

In addition to Da Lio and me, there were ten other people in the operating theater, including surgical oncologist Dr. Helena Chang, who would perform a mastectomy on the patient's cancerous breast before Da Lio and his chief resident stepped in to reconstruct the breast and perform a mastectomy and reconstruction on the patient's other side. The chief resident's name was Raquel Ulma. Da Lio told me Ulma was planning to specialize in pediatric craniofacial surgery, operating on children born with deformities, for example, and had already been training for eleven years. A visiting doctor from China also stood nearby, there to learn about American surgical techniques. Also in the room were a medical student from Sweden, at UCLA for three weeks as part of a traveling fellowship program in the United States, and a man named Andrew, who projected an unmistakable air of authority. Andrew was the operating room's head nurse. "Don't touch the tables," he told me. "Really, don't even brush against them."

The tables and the instruments on top of them were sterile. I was not. The surgical paraphernalia on each table included a stack of about fifty clamps, retractors that looked like back scratchers, gauze pads, syringes, plastic bowls and bulbous rubber suction devices that reminded me of tiny turkey basters. Trying to stay out of the way, I sat down on one of several stools scattered around the room just as the patient, a fiftysomething woman, was wheeled in. She was awake and her gray hair was tucked under a head covering that looked like a shower cap. Her hands were folded atop a mountain of blankets. Andrew stepped behind the patient's head and put his hands on her shoulders. "I'm Andrew. I'm one of the nurses who's going to be helping you today," he told her as another nurse lowered the side rails on her gurney. The patient scooted herself onto the operating table, a narrow platform just wide enough for her body with small thinner platforms jutting out from both sides. Andrew arranged a pillow under the woman's legs—"For your comfort," he told her—and secured a white strap across her thighs.

An anesthesiologist placed a clear plastic mask over the patient's face. "Some oxygen," he said. But almost immediately, it became clear the woman was getting more than just oxygen. Her blinking slowed, and within seconds her eyes closed and she was asleep. A nurse-anesthetist promptly tipped the woman's face back and up, then threaded a plastic tube down her trachea. The nurse-anesthetist taped the tube in place and stretched another strip of tape over the closed eyes. Meanwhile, the blankets that covered her upper body had been removed, leaving her naked from the chest up. I noticed that she had a lot of freckles. Her arms were strapped to the small platforms on either side of the operating table, arranging her body into the shape I remembered from my own surgery.

Da Lio stepped up to the patient, got out a purple marker, and inked over the lines on the patient's chest he had sketched earlier. Above and below each nipple, Da Lio drew a curved horizontal line, making a shape that looked like an eye on each breast. The skin inside the lines, including the nipples, would be removed and the edges sutured closed. Artificial nipples would be tattooed on later.

Chang, the surgical oncologist, stepped forward. Born in Taiwan, Chang has a doctorate in molecular biology from Temple University, where she also attended medical school. Chang is director of UCLA's breast center, a position she has held since she replaced book author Dr. Susan Love in 1997.

Before the surgery began, Chang injected a radioactive tracer into the patient's cancerous left breast. The substance coursed through the breast's lymphatic system and settled into its sentinel nodes, which were rendered temporarily radioactive. This would make them easier to find and remove during surgery. Chang waved a wand called a gamma probe over the area. A sort of Geiger counter for radiation, the wand was attached to a small machine that beeped when it detected radioactivity.

After determining the approximate location of the sentinel nodes, Chang stepped back from the operating table to make room for Da Lio. The two frequently operate together and barely speak to each other during surgery because their movements are so choreographed. As if moving a sleeping baby, Da Lio reached his arms under the patient and gently repositioned her on the table. Then Ulma, the plastic surgery chief resident, announced the details for the day's surgery: a bilateral mastectomy, a sentinel-node biopsy and a reconstruction using tissue expanders, saline-filled bags that would be placed inside the patient's chest temporarily and swapped out for implants months later.

When plastic surgeons started routinely using implants in breast cancer reconstructive surgery in the 1970s and 1980s, they placed the silicone or saline forms directly under a patient's skin. Because so much skin was removed during mastectomies, these surgeons had little space in which to work. In line with the less-is-more trend in breast cancer surgery today, many doctors leave nearly all of a mastectomy patient's skin in place, scooping out only the tissue inside. Called a skin-sparing mastectomy, the procedure allows plastic surgeons to put an implant in the same pocket that once held a patient's natural breast tissue. It's often not a perfect swap, though, and so many plastic surgeons use tissue expanders. The expanders are placed in a woman's chest directly after a mastectomy and contain a port through which saline can be injected in several in-office sessions after surgery. The saline causes the expanders to get larger, gradually stretching a patient's skin and muscle so a real implant can be placed there in a second surgery.

Ulma read the name and birth date of the patient and noted that she had recently had a respiratory infection and had no known allergies to medication. One by one, the doctors and nurses who would be directly involved in the surgery left the room to scrub their hands in a stainless-steel sink outside the door. Upon reentering, they held their wet hands in the air and were handed sterile blue towels. After drying off, they slid their arms into sterile gowns held up by nurses and had sterile gloves snapped over their hands and cuffs.

Ulma washed the patient's upper body, and the antiseptic soap she was using mixed with Da Lio's pen markings and turned into a light blue foam. The unconscious patient's flesh jiggled with each scrub. Her head rocked back and forth with the motion. After applying a second antiseptic wash, Da Lio draped layers of blue sterile cloths and sheets across the patient's body. When he was finished, all that remained exposed was a rectangle about eighteen by twenty-four inches on the woman's chest. Da Lio and Chang picked up a scalpel each and, simultaneously, cut along the lines of the eyes drawn onto the patient's breasts. As the skin separated, I could see yellow, bumpy fat beneath. The doctors handed the scalpels to a pair of nurses and each picked up an electrosurgical knife they would use for the rest of the surgery. The device, named a Bovie knife after the doctor who invented it in the 1920s, cut and cauterized in the same motion using hot electric pulses. The smell of burning flesh was unmistakable as the knife hissed and small whiffs of smoke floated upward into the operating room's air-conditioning system.

Soon, Chang had her hands deep inside the patient's left breast skin pocket, pulling out more and more flesh as she worked.

On the other side of the patient, Da Lio was further along. "We have the right breast," Ulma announced. The blob of flesh was placed on the patient's drape-covered stomach, the intact nipple still attached. I glanced at the surgery wound on her chest and when I looked back, the blob had been deposited into a small plastic bucket covered with a sealed top. The flesh would be analyzed by a pathologist for signs of cancer. Minutes later, Chang had removed the left breast as well, leaving the patient's chest flat. Inside the holes cut by the surgeons, I could see charred flesh and the woman's deep red pectoral muscles. They looked like raw steak. Chang picked up the gamma-probe wand again and waved it across the patient's armpit. When it beeped, she extracted the corresponding lymph nodes, getting to them through the same incision she had made with her scalpel. Each node she took out was labeled with a small tag. When she finished, Chang removed her sterile gloves and walked to a computer in the corner. She reviewed something on the screen, typed a few lines and, just like that, her part of the surgery was over and she left the room along with the members of her surgical oncology team.

Now it was time to build some breasts. Employing a technique that is now common across the country, Da Lio and Ulma would use the patient's pectoral muscles and strips of commercially available cadaver tissue to create internal slings for the expanders. Called AlloDerm, the cadaver tissue is specially processed to remove cells from the donor, leaving a matrix of collagen. Over time, cells from the breast cancer patient grow into the matrix, which literally becomes part of her body. In ten days, Da Lio told me, it would have its own blood supply.

The first expander was already soaking in a dish of triple-antibiotic solution. The nurse removed the AlloDerm from its packaging; it looked like a small sheet of white plastic. Da Lio sewed the AlloDerm strip to the patient's pectoral muscle using sutures that would eventually dissolve. While Da Lio worked, he asked the visiting medical student how to say Merry Christmas in Swedish and mentioned that the student's home country's parental-leave laws are legendary. As he and Ulma, the chief resident, took turns stitching and arranging the expander,

I mentioned that they seemed very relaxed. "I'm bored," said Da Lio. "Boring surgeries are good." A drama-free operating room is a safe operating room. In their two decades working together, Chang and Da Lio have performed more than five hundred reconstructive surgeries on breast cancer patients.

As Da Lio stitched the first incision closed, Ulma moved to the patient's other side to sew in a new strip of AlloDerm and situate an expander. Two other plastic surgery residents, more junior than Ulma, swung open the operating-room door. They were there to update Da Lio on the team's patients in the hospital that day. One woman had been discharged, they said. Another had been switched to a different painkiller after surgery. Even though the residents were wearing facemasks, I recognized one of them. She had helped care for me after the surgery to swap my expanders for implants.

It had been about five hours since the surgery began and my back was aching when Da Lio told me they were done. Where the patient's breasts had been, there were now small mounds and two straight lines. The edges of the eyes Da Lio had drawn on earlier had been pulled together and sewn shut. Da Lio and Ulma wrapped the woman's torso in gauze and a specially designed mastectomy bra made of stretchy fabric. The nurse-anesthetist, meanwhile, fiddled with a machine connected to the tube that had been inserted into the patient's throat. The woman began to stir, although her eyes were still taped shut. "The surgery's aaalll done," the nurse-anesthetist said. "All is well. You're just waking up. Take deep breaths in and out." I caught a glimpse of the woman's toenails, which were painted dark red. She flexed her feet as the nurse-anesthetist smoothly pulled the tube from her throat.

Da Lio and the nurse moved the patient onto a gurney and wheeled her out of the operating room and down a hallway to a recovery room. The woman was blinking slowly. I could see her smiling through the oxygen mask over her mouth. "She's here, but she's not here," said Da Lio. "She won't remember any of this." As she had been when she arrived in the operating room, the woman was covered with a mountain of blankets.

In his office one day, Da Lio told me about a young breast cancer patient he'd operated on years ago. After her surgery, Da Lio and the woman became friends and were so close that she invited him to Thanksgiving dinner with her family. Years later, the woman's cancer recurred and she had to undergo surgery again. "When I re-operated on her, to do her second reconstruction, she was almost like my sister," Da Lio said. In the operating room, the two chatted briefly before the woman was anesthetized and covered with sterile surgical drapes. "She was asleep, and all of a sudden, she stopped being my friend and became an anatomic project that had to get done," he said. After the surgery was complete, Da Lio went with the woman to the recovery room. "She woke up; I was holding her hand and she became my friend again," he told me. "Surgeons have a way of detaching. We're not trained to operate on people. We are trained to operate on anatomy."

One of the keys to being a good surgeon, Da Lio said, is repetition. "Excellence is a habit, not an act," he said. "If all you do is microsurgical breast reconstruction day in and day out, five days a week, you become very, very, very good at it."

Surgeons who regularly or exclusively perform breast cancer reconstructive surgery often have better outcomes than surgeons who primarily do other types of plastic surgery. In an academic medical center like UCLA, this kind of deep experience is easy to come by. But in smaller hospitals where most American breast cancer patients are treated and where such treatment is a small part of the job for surgical oncologists and plastic surgeons, research has shown that patients are more likely to experience complications. Frank DellaCroce knows this because he sees it every day.

DellaCroce is a plastic surgeon and cofounder of the Center for Restorative Breast Surgery, a hospital in New Orleans that caters exclusively to breast cancer patients. DellaCroce and his business partner, Scott Sullivan, who is also a plastic surgeon, are the principal owners of the facility. With twenty-three private rooms and seven operating rooms, the center has become a destination for breast cancer patients across the country and around the world. Two-thirds of its patients are from outside Louisiana, according to Sullivan. About half of the women who undergo surgery at the facility have already been operated on by plastic surgeons elsewhere but travel to New Orleans because they are unhappy with their outcomes. Sometimes, the women need new sets of implants or minor surgery to correct scars or sagging skin. Other times, women need "a ground-up rebuild," DellaCroce told me. "Some of these ladies—this is going to sound crazy—but they've been through six or ten operations before they get here. So it's been a real hard journey. You've really got one good shot to help them and get it right."

DellaCroce and Sullivan opened their hospital in 2009, four years after Hurricane Katrina devastated New Orleans. They had purchased a small space before the storm, and when it hit they evacuated along with much of the city. Sullivan lost his house in the flood. DellaCroce temporarily relocated to his hometown in north Louisiana. "I was sitting in front of the TV, watching CNN and all the craziness. And something just went off in my head and I decided to call my real estate agent," he told me. Next to the space DellaCroce and Sullivan had already purchased was an Italian restaurant. *When the flood receded, could they take over that space and build an even larger facility?* DellaCroce wondered. He asked his real estate agent, who called the restaurant owner. "For cents on the dollar, we made an offer. He accepted it and that's how it all ended up coming to fruition," DellaCroce said. After also purchasing a neighboring property that had previously been home to a Dunkin' Donuts, Sullivan and DellaCroce built the sixty-thousand-square-foot center on St. Charles Avenue in the heart of New Orleans. When I first spoke to DellaCroce in January 2018, he told me that surgeons at the facility performed some six hundred breast cancer reconstruction operations a year. When we chatted again a year later, he said the hospital's business was booming and on track to perform a thousand procedures annually.

The hospital, which accepts some insurance plans, prides itself on creating a spa-like atmosphere for its patients. Women are treated to aromatherapy, and the light-filled lobby, with its vaulted glass ceiling, is decorated with overstuffed couches and orchids. Because DellaCroce and Sullivan own the facility, they have total control over the nurses, staff and finances. "It's not a money-printing machine," DellaCroce said, but the hospital generates a reliable profit.

DellaCroce performs breast-implant surgeries similar to the one I observed at UCLA, but he and the five other surgeons working at the center heavily advertise their ability to perform autologous breast reconstructions. These surgeries are more expensive, take longer and require longer recovery times, but they create breasts that have a softer and more natural feel than implants, which must be replaced every ten years or so. Plus, when flesh is harvested from a woman's abdomen, it's akin to getting a tummy tuck, an ancillary benefit.

DellaCroce and Sullivan told me their hospital was flooded with calls in 2013 after the actress Angelina Jolie published an op-ed in the *New York Times* revealing that she'd tested positive for the *BRCA1* gene and had opted to have a preventive double mastectomy with implant reconstruction. "You see a plastic surgeon and they tell you they can do a lovely job and you'll look beautiful, but this gave it a high-profile face," DellaCroce said. Women getting preventive mastectomies represent about 15 percent of patients at the Center for Restorative Breast Surgery. Such treatment, frequently performed on young women, has raised the bar for plastic surgeons and led to improvements in techniques and outcomes for all breast cancer patients. In addition to preserving a woman's skin, many surgeons who perform mastectomies, preventive or otherwise, also now leave her nipples intact and in place. Of such surgeries performed on women hoping to prevent breast cancer, DellaCroce said, "You're taking a lovely untouched body without any known disease, and now you have to do a mastectomy on this person, and the burden on the plastic surgeon is to give them something that makes that decision tenable." In theory, a *BRCA*-positive woman will be more likely to opt for the lifesaving surgery if she knows she'll still have attractive, if artificial, breasts.

Partially in response to the needs of *BRCA*-positive women and an increase in the rate of breast cancer patients choosing to have reconstructive surgery, a sort of breast cancer plastic surgery cottage industry has sprung up in the past decade. Surgical centers that offer only breast reconstruction now exist in several major cities. Even outside these specialized centers, plastic surgery for breast cancer patients is big business. While costs (mostly reimbursed by insurance) vary geographically and by provider, reconstructive surgeries can run more than thirty thousand dollars each, and that's not including subsequent operations to replace worn-out implants or fix mistakes.

The foundation of this sector of the health-care industry is the 1998 federal law that requires insurers to cover the costs of nearly any breast-cancer reconstructive surgery, including surgery on a noncancerous breast to make a woman's chest symmetrical after treatment. In New York State, the law goes even further. In 2011, the state adopted a statute, first proposed by a New York plastic surgeon, requiring doctors to discuss reconstructive options with breast cancer surgery patients. After the law was implemented, rates of the procedures increased significantly, particularly among low-income women and women of color.

Interestingly, reconstruction rates vary widely by location. A 2014 study published in the *Journal of Clinical Oncology* found that 18 percent of breast cancer patients surveyed in North Dakota opted to have reconstruction; in Washington, DC, it was 80 percent. The study's authors attributed the state variation to the number of plastic surgeons living and working there. Women with higher incomes living along the coasts or in large cities are more likely to have ready access to surgeons trained to perform breast reconstruction.

The operations, while optional, are not without risk. A 2018 study published in the journal *JAMA Surgery* found that one-third of breast reconstruction patients experience complications in the first two years, with many requiring additional surgery.

Not all breast cancer patients choose to have reconstructive surgery. Some women choose to embrace their battle-scarred bodies or would rather not endure the painful recovery process that follows reconstruction. Every patient is unique, of course, but research has shown that reconstructive surgery improves the psychological well-being of many breast cancer patients. Sheri Weitz, the first patient to enroll in the Kadcyla trial at UCLA, had autologous breast reconstructive surgery in which tissue from her abdomen was moved to her chest. She told me that after the surgery, she was unable to stand upright for nearly six weeks, but today she has a better figure than before her diagnosis. "I look good in a bikini now. My stomach is flat. That part's good. I had my [noncancerous breast] lifted to match, so that's good," she said. Da Lio told me he can tell if a woman is particularly pleased with her breast reconstruction results: she shows up to an exam with tan lines in the shape of a bathing suit.

11

so meta

At the baggage claim in San Antonio, there were chauffeurs everywhere. They stood still and silent in black suits and button-down shirts, holding paper signs with the names of the travelers they had been sent to fetch. I grabbed my small purple suitcase from the conveyor belt and watched as the suited men were engulfed by small groups who then piled into cars waiting outside. Many had come a long way, from Saudi Arabia and Spain and Taiwan, among other places.

There was no chauffeur for me, just a Lyft driver named Ismael who listed his favorite local restaurants as we snaked along U.S. Route 281 south toward downtown San Antonio. "You like burgers? You gotta try Whataburger," Ismael insisted. It was long after dark by the time I checked in to the Hilton Palacio del Rio, clicked the hotel-room door shut, flopped my suitcase onto the bed and slipped off my shoes. I walked across the room and pulled open the curtains so I could step out onto the balcony. I had an unobstructed view of the reason I had come to Texas.

The Henry B. González Convention Center was closed, but the lights were on, and they emitted a yellowish glow amid the green and red twinkling lights strewn along the San Antonio River for the Christmas season. Throughout the week, more than seven thousand doctors, scientists and pharmaceutical company representatives would descend on the city for the world's largest annual scientific breast cancer conference. The foreign travelers I had seen at the airport baggage claim were on their way to the conference. I was supposedly in San Antonio to conduct some interviews and get a handle on the latest in breast cancer science. In truth, though, my dream date was not with any researcher or doctor. The attendees I was most eager to meet didn't have medical degrees or PowerPoint slides or seats on any dais.

To reach them the next day, I had to wade through a sea of commerce. Inside the convention center's main exhibition hall, it seemed that every major pharmaceutical company was putting on its best come-hither show for conference attendees. A pair of young, lithe dancers whipped flowing fabric through the air at an exhibition booth for the drug Faslodex, a new injectable from AstraZeneca used to treat women with estrogen-fueled advanced breast cancer. Novartis had free cupcakes. Tesaro, a company developing new drugs for *BRCA*-linked breast cancer, had free Nutella-branded ice cream cones. Espresso was available at Eli Lilly, and Pfizer had put out small cups of frozen yogurt. Takers could choose their toppings, purple sprinkles or blueberries, to match the packaging for Ibrance, the CDK4/6 inhibitor Dennis Slamon had helped bring to market. Medtronic, a medical-device company, had breasts of raw chicken at its booth so surgeons could test the PlasmaBlade, the company's new soft-tissue-dissection knife.

At the back of the exhibition hall, past the plush carpet squares and free food, I found the table I was looking for. There was no complimentary coffee or ice cream, just brochures stacked in neat rows and a small sign that said METAvivor. Founded in 2009, METAvivor is a nonprofit organization run by and advocating for metastatic breast cancer patients. In recent years, the group and others like it have tried to turn public attention to the more than one hundred and fifty thousand women in America living with stage IV breast cancer. For decades, the needs of such women have been overshadowed by a pink-ribbon culture focused on "survivors." Having a presence at major breast cancer conferences is part of a multipronged strategy to increase research funding for metastatic disease and raise awareness that for all the strides made in treatment, some forty thousand American women still die every year. Women dying of breast cancer don't elicit the kind of feel-good philanthropy that has helped fund research in screening or curing early-stage breast cancer. And so metastatic breast cancer groups have had to get creative.

One group, called METUP, founded in 2015, stages public "die-ins" at cancer conferences and on a lawn next to the U.S. Capitol building. METUP activists are sometimes arrested during protests and frequently criticize the efforts of traditional breast cancer–advocacy groups like the Susan G. Komen foundation. The group modeled its nonviolent, civil-disobedience tactics on those of ACT UP, the activist group that drew the world's attention to the AIDS crisis. In a 1988 protest, ACT UP demonstrators blocked and eventually closed the headquarters of the Food and Drug Administration in Rockville, Maryland, to spur the agency to approve HIV and AIDS drugs more quickly. Susan Rahn, a metastatic breast cancer patient and president of METUP, frequently posts anger-laden messages on Twitter accusing Komen and other mainstream advocacy groups of not devoting enough resources to stage IV disease. Some of the Tweets contain profanity, and Rahn told me that's intentional. "I get on Twitter and I rage," she said. When plotting her Tweets, Rahn thinks, "What are we going to do to Komen? How are we going to go after the National Institutes of Health? How are we going to go after the FDA? Not just those institutions, but the actual people there. We need to get to those people and make them feel it." METUP's fund-raising efforts go toward logistical support for its high-profile activism. METAvivor, however, raises sorely needed money to support research on metastatic breast cancer.

Despite the billions of dollars collected and spent on breast cancer research over the past half century, relatively little has been devoted to studying metastatic breast cancer patients or their particular forms of the disease. Doctors do not know why some breast cancers eventually form deadly metastases or how to permanently quash the disease once it has spread. Patients with metastatic disease are typically treated with one drug after another, their doctors switching the medications whenever the disease stops responding to treatment. (In some cases, they are also treated with radiation or surgery.) Eventually, nearly all patients with metastases run out of drugs and die, although in recent years many have been living longer. The five-year survival rate for women under fifty who already had metastatic breast cancer at their initial diagnosis increased from 18 percent between 1992 and 1994 to 36 percent between 2005 and 2012, according to the National Cancer Institute. This is due primarily to new drugs, including Herceptin. As metastatic patients are living longer, they are organizing themselves into a force powerful enough to change the funding priorities of non-profit groups like Komen. Of the four metastatic breast cancer patients who founded METAvivor in 2009, three have died, but one is still alive, thirteen years after she was diagnosed with stage IV disease.

In June 2018, researchers at the National Institutes of Health reported that a metastatic breast cancer patient appeared to have been cured of her disease through an experimental immunotherapy treatment. But such stories are incredibly rare. Breast cancer that spreads most commonly finds its way to a woman's bones, brain, lungs, liver, or several of these sites. The latter three can be fatal, as the disease impedes the function of these vital organs.

Three-quarters of women with metastatic breast cancer were originally diagnosed with early-stage disease. The idea that the breast cancer "came back" after initial treatment is a bit misleading. Women who undergo traditional chemotherapy or other drug treatments shortly after an early-stage diagnosis get these treatments because their doctors believe they have micrometastases—cancer growths outside the breast that are too small to appear on scans. Breast cancer that is treated and later metastasizes into larger tumors around the body somehow managed to survive the initial onslaught of surgery, radiation and drugs. Women who declare themselves cancer-free after treatment have no way of knowing if that is true. Sometimes breast cancer continues to grow during treatment. Other times, breast cancer cells that remain alive after treatment go dormant but then begin multiplying years or even decades later. Many women treated for early-stage breast cancer may believe they are cured, but the disease can and does recur in some one-third of patients. It's impossible to predict with certainty which women will fall into this group.

Lianne Kraemer, a smiling, energetic woman with dark brown hair and eyes, had been living with metastatic breast cancer for more than a year when I met her at the METAvivor booth in San Antonio. Diagnosed with estrogen-receptor-positive breast cancer in January 2014 at the age of thirty-seven, Kraemer had a double mastectomy, chemotherapy, lymph-node removal, radiation and hormone treatment. She emerged from the yearlong ordeal relieved she had acted quickly—she'd felt a lump in her breast and was diagnosed the next month. Then, one day in June 2016, Kraemer was putting a load of laundry into her washing machine when she suddenly felt an unfamiliar tingling in the right side of her lips. The next day, the right side of her gums and tongue started to feel weird too. "During that day, the crease on my right index finger felt like someone had put a slight rubber band around it," Kraemer told me. Kraemer was a speech pathologist who treated children with disabilities, so she was accustomed to evaluating neurological problems. She knew something serious could be happening inside her brain. But the idea that it could be related to her breast cancer? It was possible, sure, but unlikely, she thought.

Kraemer e-mailed her oncologist, who recommended she get a brain MRI. Kraemer is from a close-knit family in St. Louis and she asked her uncle, a neurologist, to review the scan. She had been a flower girl at his wedding. When Kraemer's uncle broke the news that it looked like her breast cancer had spread and formed about a dozen small tumors inside her brain,

Kraemer was stunned. "Everything just went very quickly from there," she said.

Kraemer's oncologist immediately ordered scans of the rest of her body to assess the disease's spread. It is extraordinarily rare for a patient with metastatic estrogen-receptor-positive breast cancer to have the disease migrate to her brain and nowhere else. But this was what had happened. The scans showed no signs of cancer elsewhere in Kraemer's body. That Kraemer had brain-only metastases provided a possible clue about how she went from being an early-stage breast cancer patient to a woman with a terminal disease. It's likely Kraemer's breast cancer had already spread to her brain by the time she was diagnosed. Any errant breast cancer cells in other parts of her body had been killed off by the chemotherapy, but the chemo had not reached Kraemer's brain because of what's known as the blood-brain barrier—tight-knit capillaries that prevent almost all large molecules (including most chemotherapy drugs) from leaving the blood and entering the brain tissue. The blood-brain barrier protects the brain from harmful elements, but in Kraemer's case it will probably cost her her life.

Often, although not always, newly diagnosed breast cancer patients will get abdominal scans, but they almost never have brain scans unless a symptom appears. (About a year after I finished treatment, I had a debilitating two-day headache and my oncologist recommended I get a brain MRI, which turned up nothing. I was diagnosed with a migraine.) Even if Kraemer had had a brain scan when she was diagnosed, it's possible the tumors inside her skull were so small that they would have been invisible on an MRI.

As Kraemer explained all of this to me in a phone call a

few months after we met in San Antonio, it was obvious she had thought through every possible diagnostic scenario over and over again. When one is diagnosed with metastatic breast cancer, there is a lot of second-guessing. Could her doctors have done more? Did the month that passed between the time Kraemer found a lump in her breast and her diagnosis make a difference? Kraemer's treatment started with a double mastectomy, and she had surgical complications that delayed the start of her chemotherapy. Did this give the cancer time to spread to her brain? She will never know but she told me that most likely, her metastatic breast cancer could not have been prevented. "You can do everything right and still end up metastatic," Kraemer said. "I just assumed that could never be me because I would catch it early, because I was on top of things. I believed the narrative that is pushed on women, that if you check your breasts and if you catch it early, you're fine. That surely the women who are not okay must not have gotten good treatment or must not have caught it when they should have. I believed what I was told."

Existing breast cancer–treatment protocols fail in tens of thousands of women like Lianne Kraemer every year. In some cases, breast cancer patients don't have access to high-quality treatment or they ignore signs of the disease until it's incurable. But more women who die of breast cancer succumb to the disease for no other reason than that it manages to outwit the protocols.

Pharmaceutical companies and researchers often test new drugs on metastatic patients before anyone else. These are women who are dying anyway and they are the ones most willing to be part of experiments. But the goal for most new drug development is to treat early-stage patients successfully and eradicate breast cancer before it becomes metastatic. There is little focus on studying metastatic disease itself. In a way, this is understandable. The work is incredibly difficult. Cancer is programmed to stay alive. When cancer cells are hit with drugs, they often mutate to survive. As the cells develop immunity to more and more drugs, they become stronger, more unique and more complex.

Cyrus Ghajar is one of a relatively small number of scientists studying metastatic breast cancer full-time. A cancer biologist at the Fred Hutchinson Cancer Research Center in Seattle, Ghajar is particularly interested in why some cancer cells remain dormant for years and then grow into metastatic tumors. Ghajar and the other scientists working in his lab have received millions of dollars in grants from the federal government and nonprofit research groups like Komen and the Breast Cancer Research Foundation.

When I reached Ghajar by phone to ask about his research, he told me so many fascinating things about metastatic breast cancer that the need for further study was immediately obvious. Ghajar told me, for instance, that 25 to 40 percent of early-stage breast cancer patients already have cancer cells in their bone marrow and that these patients are, on average, three times more likely than their counterparts to develop metastases later. It's impossible to know if a patient has cancer cells in her bone marrow without driving a long needle into a large bone and aspirating marrow from inside. Ghajar said that, in theory, patients could give consent to have their marrow aspirated when they are placed under general anesthesia for lumpectomies or mastectomies. These aspirations could tell doctors which women were more likely to face a recurrence of their breast cancer and so might need more treatment up-front. Sampling could also help researchers to study the cells and microenvironment of the bone marrow to learn more about why some women with cancer cells in their marrow at diagnosis later develop metastases and why others do not. "We don't have a way to further stratify people, because we haven't studied it enough. We don't have enough samples," Ghajar told me. He also said that metastatic breast cancer in a woman's liver, for example, is almost certainly different than the metastatic tumor in her brain. He and his fellow scientists are trying to untangle the differences in hopes of one day refining treatment for metastatic patients.

Ghajar pointed out that the Cancer Moonshot, a National Cancer Institute initiative launched by President Barack Obama and Vice President Joe Biden, does not explicitly provide funding to address the unique challenges of metastatic cancer. "How can you have a moonshot trying to cure cancer and not mention people *dying* of cancer?" he said.

Alana Welm, a molecular biologist, runs a lab at the University of Utah's Huntsman Cancer Institute devoted to studying breast cancer metastases. One of her research projects compares breast cancer cells from a patient's original tumor against cancer cells found elsewhere in the body after the disease spreads. The idea is to find differences between the cells for clues about why some migrated and how they may then be affected by microenvironments inside the body. "Think about how hard this research is," Welm said. By the time a woman is diagnosed with metastatic disease, her original biopsy tissue may no longer be available. In addition, it's often hard to get a sample of a metastatic tumor, which may be buried inside the brain or located in a place that is difficult to access safely. "I sometimes wait for years in between to try to get these samples," Welm said. To gather more samples of metastatic breast cancer, Welm and other researchers encourage stage IV patients to consider making plans to have an autopsy done immediately after they die so that cells can be harvested and preserved before they degrade.

Research into early-stage breast cancer requires none of these extreme measures. Nearly every patient diagnosed with early-stage breast cancer has a biopsy that produces readily available tissue. Nearly all labs that test drug compounds on breast cancer cells work with early-stage samples. This means a drug developed in the lab may work against early-stage disease but might be useless for women with metastatic breast cancer.

Another challenge in studying metastatic breast cancer, Welm said, is that the lab does not replicate the environment in which the disease naturally exists. Cells living in a petri dish or grafted onto a mouse are removed from the human immune system and the unique environment where metastatic breast cancer thrives.

After I finished my treatment for early-stage disease in February 2016, I did not celebrate. I did not feel cured. I felt scared. Although it is unlikely, my disease could recur. I was unnerved when I learned that I would not get regular or even annual scans to look for signs of metastases. Studies published in the 1990s showed that detecting metastases through scheduled scans, rather than when a symptom appears, did not change survival, prognosis or quality of life. The patient would die either way. Scanning every breast cancer patient post-treatment would be expensive and would undoubtedly lead to medical intervention when it is not needed. Imaging studies often pick up abnormalities that look suspiciously like cancer but are not.

Recent research, however, has suggested that metastatic breast cancer that is smaller in volume and contains fewer mutations may be easier to treat than cancers that are larger and more complex. This has led some researchers to wonder whether detecting the spread of breast cancer earlier may have benefits. Imaging has gotten much better since the 1990s, with MRI, positron-emission tomography (PET) and CT scans able to pick up tumors as small as just a few millimeters. In 2016, well-known cancer doctor and researcher George Sledge published a paper in the *Journal of Oncology Practice* called "Curing Metastatic Breast Cancer." In it, Sledge suggested that the existing paradigm around metastatic breast cancer—that it is incurable and not worth looking for—should be updated in the face of new science developed in the past decade or so. Pointing to the fact that 1 to 2 percent of metastatic breast cancer patients are cured of their disease or survive for many years, he wrote, "If some patients are cured, might not we cure more?"

About a month after her diagnosis of metastatic breast cancer in the summer of 2016, Kraemer was going through all of her belongings in St. Louis, hoping to save her parents the agony of discarding her possessions after she died. She emptied a box, and her hands settled on a photograph she hadn't expected to find. It was a picture of a smiling couple, Kraemer and an ex-boyfriend. He had been her first love. After the man had unceremoniously dumped her more than a decade earlier, Kraemer had thrown out every physical reminder of him. He had given her a waffle iron as a present, and she smashed it to bits in an alley. But somehow, this photograph had survived the reckoning. Kraemer held the photograph in her hands for a long while, her mind wandering back to the year she and this man, a Nebraska native named Eric Marintzer, had spent together.

Like any modern woman, Kraemer picked up her iPhone and found Marintzer's Instagram profile. A talented amateur landscape photographer, he had recently posted some images from a trip to Banff National Park in Canada. The photographs of alpine lakes and mountain peaks were stunning. Kraemer scrolled through more and more images, her thumb moving so quickly on the touchscreen that she accidentally hit the Like button for one. She promptly panicked. The man who'd broken her heart would know she had looked him up. Kraemer threw her phone across the room so hard that the screen shattered. She decided to come clean. She had just been diagnosed with a terminal disease. What did she have to lose?

Kraemer sent Marintzer a message through Instagram, saying she had seen his photographs and accidentally liked one of his images. She complimented his photography and wished him well. To her surprise, he wrote back. Then they started texting. The conversation was casual, but eventually Marintzer asked Kraemer out for drinks. When she arrived at the Chicago bar where they had agreed to meet, Kraemer sat in her car until she saw Marintzer walk in. She planned to confront him. "My plan was to basically rip him a new one for the way he handled our breakup. We sat down and we were laughing. Three and a half hours later I'm like, 'I have to go,'" Kraemer said. She told Marintzer that she had been treated for breast cancer. She didn't mention that it had spread to her brain.

Kraemer was living in St. Louis with her family but often traveled to Chicago, where Marintzer lived and where they had met. Whenever she went, they got together. It was refreshing to be with someone who didn't know about her disease. "Everybody in my life was treating me with kid gloves," Kraemer said. The two grew so close that they spent New Year's Eve together. That night and the next day, they talked about whether they should officially be a couple again. "There's something you need to know," Kraemer finally said. She told him everything. He listened and then said her metastatic breast cancer was no reason to change their plans. Before long, Kraemer moved into Marintzer's condo and they picked up where they had left off as twentysomethings, albeit with different challenges to confront.

Meanwhile, Kraemer tried to stay alive. Women who have breast cancer that has spread to the brain often die within a year or two of the diagnosis. To shrink the tumors inside her brain, one oncologist recommended she undergo what's known as "whole brain radiation." The treatment, an extreme but common procedure for patients with multiple brain tumors, can cause debilitating fatigue. More worrisome to Kraemer was that it could leave her with permanent cognitive damage and possibly some form of paralysis. Another oncologist proposed that Kraemer enroll in a clinical trial testing whether a CDK4/6 inhibitor, Verzenio, could beat back brain tumors in women with estrogen-receptor-positive metastatic breast cancer. Kraemer went on the drug, and most of her tumors stayed the same size for eight months, a good sign. The largest tumor, the one that had caused her face and hand to tingle, even shrank. The drug also kicked Kraemer into chemical menopause, which her oncologist thought might help cut off the supply of estrogen her type of breast cancer was feeding on.

But the drug had side effects, including chronic diarrhea that required Kraemer to get IV fluids to prevent dehydration. Her oncologist dropped the dose of the trial drug, which stopped the diarrhea but also the menopause. Soon after, her brain tumors began growing again. The clinical trial protocol dictated that Kraemer had to stop taking Verzenio. If her cancer was growing, the drug was not working. But here's where it got interesting. Kraemer switched to a new oncologist, Dr. Nancy Lin, a metastatic specialist at the Dana-Farber Cancer Institute in Boston. Lin speculated that the drug actually was working but that the estrogen produced by Kraemer's ovaries was canceling out its effects. (Kraemer had taken tamoxifen after her initial diagnosis in 2014 but stopped in 2016 due to side effects; she had been about to switch to an aromatase inhibitor when she was diagnosed with brain metastases.) The clinical trial rules did not allow patients to start hormone therapy in the middle of the study, as it could muddy the results. To Lin and Kraemer, this was unacceptable. "I said, 'Listen, I need the [Verzenio] back. I don't care what I have to sell or do, someone's going to get me that damn drug back,'" Kraemer told me. Lin lobbied the drugmaker, Eli Lilly, to allow Kraemer to have Verzenio and begin hormone therapy. Miraculously, the company agreed. It would sponsor a new trial that included just one patient, Kraemer.

The hormone-therapy-and-Verzenio combination, which is now a standard treatment regimen, kept Kraemer's brain tumors stable for six months. In January 2018, a scan showed they were growing again and Kraemer was advised by a radiation oncologist at Dana-Farber to undergo whole brain radiation. But the cognitive side effects worried her, so she got a second opinion from a radiation oncologist at the University of North Carolina, Chapel Hill, who recommended a specialized radiation treatment that would hit only her largest tumor. Her brain tumors remained stable for a few months, but eventually they started growing again. In the hands of a research oncologist like Lin, Kraemer was able to enroll in yet another clinical trial, this one testing whether a drug that has been shown to work against certain types of liver and kidney cancer might help women with metastatic breast cancer.

As I write this, Kraemer's disease is at its most threatening. A brain scan in January 2019 showed that several of Kraemer's tumors were growing again, which meant she had to stop taking her latest experimental drug. She is running out of options, and the options are increasingly unappealing. She is currently debating whether to have whole brain radiation. There is an older small-molecule type of chemotherapy she hasn't yet tried that could be her next and last drug. Lin wonders if an experimental new antibody-drug conjugate, similar in design to Kadcyla, might keep Kraemer's cancer at bay a bit longer, but it's available only through a clinical trial. Lin has been unable to convince the drug's manufacturer that breast cancer patients with brain metastases should be allowed to enroll. Drugmakers often exclude patients with brain metastases from clinical trials, fearing their typically poor outcomes might dampen results.

In September 2018, before Kraemer received the news about her tumors growing again, I dug out the winter clothes I never need in Los Angeles and flew to Chicago to see her. We met at a charming, upscale diner for breakfast the day after I arrived. Kraemer looked as beautiful as ever. When she walked into the restaurant, her face, with its high cheekbones and smooth skin, was flushed from the cold. Marintzer trailed behind her. As the couple settled into a booth and I slid in across from them, I couldn't help thinking they looked entirely, astonishingly normal, their predicament and fate hidden inside a relationship they were trying hard not to define by the fact that Kraemer was battling a disease that would most likely cut her life short.

As we talked and ate our breakfast, I asked Marintzer how he had felt when he learned that Kraemer had terminal cancer. "There's a healthy chunk of denial—just enough to keep you steady," he said. "But I'm not stupid. We know what's going to happen with Lianne." Plus, he joked, by the time he learned the truth about Kraemer's cancer, "She had stuff at my house. I'd fallen in love with her dog. She had already sunk her claws into me."

In the two years since they rekindled their relationship, Kraemer and Marintzer have traveled to Italy, Ireland, Colorado and northern Michigan on vacation. Kraemer said that shortly after her diagnosis she had been understandably distraught. "I couldn't look at my nephews without bawling my eyes out—the thought of them growing up and not knowing who I was and how much I loved them. And what would holidays look like without me? Would they forget about me eventually?" she said. With Marintzer, she told me, "It's easy to plan. He forces me to be present and moving forward."

Later that day, Kraemer drove me to the suburban therapy center where she had worked as a speech pathologist for fifteen years before she had to stop due to her treatment. She wanted me to get a glimpse of her old life. In the car on the way there, I asked Kraemer how her disease affects her everyday life. She said the largest tumor in her brain has begun to impede her ability to use her right hand. She can no longer hold chopsticks or write legibly. She also has frequent migraines and once burned herself on a hot pan because she couldn't feel that her hand was on it. She said ever since her targeted brain radiation, she has had trouble multitasking and struggles to have a conversation if music is playing in the background.

In our first phone conversation in early 2018, Kraemer had told me that, in a way, she had been happier since her diagnosis than she was before it. "I was thirty-seven and unmarried, and that certainly weighed heavily on me," she said. "The societal pressures that are put on you of what you should or should not be doing—I don't have those anymore. Truly, I have the freedom to do whatever I want to do and I really don't care, for the most part, what people think about me or my decisions." I asked her in the car if she felt like she was dying. "Yes and no," she said. She had a loving boyfriend, a tight-knit family and a good life, she said. But her symptoms were continually getting worse, which was disconcerting. "I am in a decline," she said.

As we approached the treatment center, we passed a school where Kraemer used to watch cheerleaders practice on the lawn. "This is weird," she said. "For fifteen years, I would drive by this school on my way to work." In the parking lot of the center, Kraemer craned her neck to look at the cars, wondering if she would recognize any. We walked inside the facility, where children with a range of disabilities get various types of therapy six days a week. Before long, we spotted a group of Kraemer's former coworkers in a break room. They squealed when they saw her and formed a line to hug her and catch up. One therapist's daughter had recently started college. Another had a new baby. One woman asked how Kraemer was feeling. "I'm aboveground!" she joked. No one laughed.

After we chatted with the coworkers, Kraemer took me on a tour of the center, past rooms with treadmills, yoga balls and other equipment used for physical therapy. In a hallway reserved for speech therapy, she showed me stations where parents could look through windows and observe their children being treated. Eventually, Kraemer steered me toward her old therapy room, where she'd helped children with their speech and with motor skills required to eat. A new therapist now occupied the room, but she was not working that day. Kraemer began opening cupboards, which were full of toys used for therapy. She furiously picked up one toy after another and turned each over. Unmistakably, on several toys, there it was, *Lianne,* written on the bottom. She pointed out laminated signs she had made years earlier that still hung in the office. "I thought I would be weepy and I'm not," she said. "It's very nostalgic how everything has changed, but nothing has changed at all. I'm not here anymore, but yet I still am."

On our drive back to the city, Kraemer and I talked about loss. "All I wanted to do was be a mother. It didn't work out that way and that's hard," she said. Kraemer babysat as a highschooler, worked as a nanny for twins in college and chose a career working with kids. After her early-stage diagnosis, Kraemer underwent fertility treatment and froze eggs harvested from her ovaries. Every three months, she pays to keep the eggs in storage. "I know I'm not going to use them, but I can't stop paying for them and just throw them away," she said.

Later that night, Kraemer and Marintzer invited me to their condo for pizza. I put my notebook away and we stood around their kitchen island talking about travel and our favorite television shows. After we ate, we retreated to their downstairs den and sipped wine while we talked about the current state of breast cancer research.

In December 2017, Kraemer had told me that the upcoming Christmas holiday would probably be her last. But in December 2018, Kraemer and Marintzer went shopping for a Christmas tree. Kraemer searched the lot and picked out the biggest one she could find. "It was a picture-perfect night," Marintzer told me. "Chilly, but not miserable. There was music playing and Christmas lights strung up." Marintzer and Kraemer loaded the tree on top of their car and lugged it into the condo. It was so large, they had to move most of their furniture out of their living room. "We know that the number of Christmases Lianne has are limited," Marintzer told me. "Let's do it up." For New Year's Eve, Marintzer said they were planning to eat dinner at a fancy Chicago restaurant and spring for tickets to a party at a local bar. A few months after that, they were planning a trip to Napa.

12

prognosis

Tatiana Prowell had a bright future ahead of her. After graduating from Bard College, where she studied literature and creative writing, Prowell went to medical school at Johns Hopkins University. She stayed at Hopkins for a residency in internal medicine and then a fellowship in oncology. Her mother had been diagnosed with lymphoma about a decade earlier and her father was a scientist. Prowell was looking forward to a long career as an academic cancer researcher. But then someone came along and upset her plans. An adjunct professor at Johns Hopkins who also worked for the Food and Drug Administration invited her to switch gears and work for the federal government in the agency's oncology division.

Being an oncologist at a medical center like Johns Hopkins meant having a front-row seat to the latest advances in cancer medicine. It also provided Prowell an opportunity to design and run clinical trials. But to secure funding for such studies, Prowell would have to spend a hefty chunk of her time writing grant proposals. Prowell hated the uncertainty of constantly asking for money, especially in a funding environment that tended to award money to proposals that did not take big risks or ask big questions. Being an FDA regulator offered something else, a chance to influence the entire field of oncology and help decide what drugs cancer patients could access. "When I told the faculty at the cancer center that I was going to do this, they told me point-blank that I was making a terrible mistake," Prowell said. Colleagues warned Prowell that a smart, young doctor like her could have a long and successful career as a clinical investigator and that she would be closing doors by going to the FDA.

Prowell ignored the warnings and joined the FDA as a medical officer in 2006. The adjunct professor who had offered Prowell a position at the FDA was Dr. Richard Pazdur, head of the FDA's cancer division. An innovative leader who has, in recent years, made the FDA's cancer-drug-approval process a model for other divisions of the FDA, Pazdur had previously worked as a researcher at M. D. Anderson. When he came to the FDA in 1999, Pazdur found the cancer division badly in need of reform. Regulators reviewed drug-approval applications for all types of cancer and operated in relative isolation from the pharmaceutical companies and patients affected by their decisions. In 2011, Pazdur reorganized his division into disease-specific groups so that regulators could work on drugs specifically for particular types of cancer. He also encouraged more conversation and cooperation among drugmakers, regulators and oncologists and has said he wants FDA regulators to "interact with the outside world." In line with this mission, Pazdur told Prowell she could see patients at Johns Hopkins and work part-time for the FDA, where she would review drug applications and serve as a bridge between the doctors and patients on the front lines and the pharmaceutical companies developing new medicines. Back then, such arrangements were rare, but in the past decade Pazdur has expanded the ranks of FDA regulators who are oncologists and also work in clinical practice. Of the eight FDA medical officers who work on breast cancer drugs, five, including Prowell, also see patients.

Despite all the progress that has been made in breast cancer science and treatment, the disease remains a top killer. The Susan G. Komen foundation and the National Breast Cancer Coalition are among groups that have pledged to end or dramatically reduce deaths from breast cancer in the coming years. Accomplishing this unquestionably depends on developing better drugs. In part because regulators like Prowell are more in tune with patient suffering and the failures of existing treatments, the FDA has increased the speed at which it approves new breast cancer drugs and increased the number of cancer drugs it approves every year. (Pazdur's wife, Mary, was diagnosed with ovarian cancer in 2012 and died of the disease in 2015. Pazdur has said the experience has influenced his approach to drug approval.) Under Pazdur, the FDA has also urged drugmakers to reconsider, for example, when and whether to exclude patients with brain metastases, like Lianne Kraemer, from studies. "When I would go to the clinic and see patients with a lot of unmet needs who were not able to enroll in trials, this made me review trial designs and eligibility criteria differently," Prowell said. "It gave me a sense of urgency about doing drug development as quickly as possible."

The stories of Herceptin and CDK4/6 inhibitors demonstrate that speed has not always been associated with breast cancer–drug development. And in a way, the more people learn about breast cancer, the harder it is to change the status quo. As scientists and doctors across the country better understand the variations inherent to breast cancer, the disease is sliced into smaller and smaller subtypes so treatments can be tailored in more personalized ways. This means that amassing enough patients with a specific subtype of breast cancer is more difficult and can take a long time, particularly because many patients cannot travel to the academic medical centers where trials often take place.

FDA regulators who work on breast cancer know this, and, under Pazdur, they are responding. One need only look at one of the most innovative breast cancer–drug trials currently under way to see the proof. Called I-SPY 2, the trial is overseen by regulators at the FDA and headed by Laura Esserman, the outspoken University of California, San Francisco, doctor. Unlike a stand-alone trial, in which an existing treatment is tested against a single new drug, I-SPY 2 tests multiple drugs in patients with different types of breast cancer at the same time, using a sophisticated mathematical model meant to extrapolate information about drug efficacy faster. The algorithm allows I-SPY 2 investigators to study drugs with fewer patients and report results more quickly.

In the past, proposals to conduct studies like I-SPY 2, known as platform trials, were unlikely to make it through the FDA's regulatory apparatus. Today, the FDA encourages them. In addition to testing multiple drugs with fewer patients, the FDA also allows protocols for I-SPY 2 to change in real time as the investigators involved learn more and alter the study based on new information.

"I'm completely a fan of platform trials that allow us to adapt in a more fluid fashion, enroll patients in a more efficient manner and answer questions more quickly," Prowell said. By understanding more quickly whether a drug is better than existing treatments, doctors involved in I-SPY 2 can discontinue treatments they know aren't working and switch patients to different drugs. This kind of adaptation is more ethical, Prowell says, and could help women avoid toxic drugs that don't reduce their chances of dying from breast cancer. Another aspect of I-SPY 2 that makes it stand out among breast cancer studies is that it relies exclusively on neoadjuvant therapy, in which drugs are administered to patients before they undergo surgery. (Such therapy is unusual for women with the most common types of breast cancer.) Using sophisticated MRI modeling, I-SPY 2 radiologists, like those working with Henry Kuerer of M. D. Anderson, are trying to predict pathological complete response—whether a patient's cancer has been killed off without cutting her open to find out. This also helps speed up the process. Rather than waiting to see how many patients live or die, trial investigators are determining the effectiveness of drugs based on MRIs and surgery outcomes just a few months into each study participant's treatment.

When drugs or drug combinations "graduate" from I-SPY 2, trial leaders can predict, based on mathematical models, how likely the drug is to beat existing standard treatments in final phase 3 trials, in which large numbers of patients are studied in experimental and control groups. It's an unconventional approach to drug development that I-SPY 2 leaders say should chart a course for how cancer studies are designed in the future.

"There are naysayers," admitted Donald Berry, a statistician and consultant on I-SPY 2. Some oncologists have criticized, in particular, I-SPY 2's dependence on neoadjuvant therapy and pCR rates, which produce data they say is not as reliable as overall survival rates. But, said Esserman, a different approach to clinical trials is precisely what is called for, a systemic change not just to the menu of drugs available but to the entire process of bringing them to market. "What we're trying to do is work with the FDA to change the whole paradigm," Esserman told me.

When it was over, I slipped off the blue gown and changed back into the T-shirt I had worn to the hospital. I found my brother in the waiting room where I had left him. We walked down a hallway and into an elevator, where I pushed the button marked *L*. When we got to the lobby, I walked across a tile floor and out a set of glass doors into the Southern California midday sunshine. After a few steps, I stopped. The tears that had been welling up inside my eyes were dripping down my cheeks faster than I could wipe them away. "I thought this might happen," my brother said, putting his arm around my shoulder and leading me to my car, which was parked in a space reserved for cancer patients.

I climbed into the passenger seat and the tears kept coming. I was having a real cry, one of those shoulder-heaving, gasp-inducing cries, the kind that happens when you are very sad or scared or hurt. But I was none of these things. I was, simply, relieved. The reason I was at the hospital that day was so that I could receive the last of twenty-five radiation sessions. The reason I was crying in my car was that the identity I had inhabited for nine months, the one that allowed me to leave my car in a special parking space, was suddenly gone. I still had more

HER2-specific intravenous drug treatments to endure, but those would be easy. They wouldn't make me tired and queasy like chemo or leave me doubled over in pain like surgery. The radiation burns covering the upper left quadrant of my torso would no longer get worse by the day. In a month, they would be gone. Three days after my last radiation session, I would start a new job as an assistant professor of journalism at Loyola Marymount University, a position I had interviewed for between chemotherapy sessions many months earlier.

I had gone from being a happy mom and wife to a dying woman and back again in less than a year. This book began as an attempt to understand that boomerang. Cancer had forced me to confront my own mortality and taught me to appreciate the joyful moments in life, but it had done something else as well. It had left me feeling unmoored, as though nothing could ever be predicted. With my diagnosis, the world suddenly went from a place where evidence informed future outcomes to a place where, regardless of the numbers, everything seemed binary. I would lose my breasts or I would not. I would undergo chemotherapy or I would not. I would die or I would not. The disorientation I felt when I was diagnosed did not dissipate once I finished treatment. That I would die of breast cancer was an increasingly remote possibility, but still a possibility, making my survival feel fifty/fifty even though I knew that, on paper, it was more like ninety-five/five.

I am not the first cancer patient to feel disconnected from the probabilities associated with surviving cancer. S. Lochlann Jain, a medical anthropologist at Stanford and former breast cancer patient, has written about her repeated attempts to reframe her sense of self in the face of statistics on the likelihood she would survive her disease. She calls this confusing state "living in prognosis," writing, "It's not quite that one's own survivorship is contingent on others' deaths. But the contemporary cancer discourse of survival against the odds seems to veer too far in the other direction, neglecting those in the category whose deaths have built those very odds." Jain too was disoriented and shocked by her diagnosis. "The prognosis purees the I-alive-you-dead person with the fundamental unknownness of cancer and gloops it into the general form of the aggregate... Living in prognosis severs the idea of a timeline and all the usual ways we orient ourselves in time."

After he was diagnosed with abdominal mesothelioma in 1982, Stephen Jay Gould, a Harvard paleontologist and evolutionary biologist, learned that the median survival for such patients was eight months. For a few minutes, Gould felt panicked. But then he remembered his extensive training in statistics and realized that eight months was merely the middle of a continuum charting the life spans of all abdominal mesothelioma patients. Some lived for less time, others for much longer. In his oft-cited essay "The Median Isn't the Message," Gould wrote that his advantages—relative youth, diagnosis at an early stage, access to high-quality care—meant he was likely to fall on the better side of the survival continuum. "I knew how to read the data properly and not despair," Gould wrote. Indeed, his optimism was well founded. Gould lived another twenty years before dying of an unrelated cancer.

In the summer of 2018, I knocked on Elisa Long's door looking for some wisdom about numbers. Long, a statistics professor at UCLA, had been diagnosed with triple-negative breast cancer at thirty-three. When we met, she was two months pregnant with her first child, a result of fertility treatments she had undergone during her cancer treatment and before she had her ovaries surgically removed to reduce the chances her disease would recur. Long's diagnosis was about as improbable as mine and her prognosis was, like mine, excellent. She was unlikely to experience breast cancer again, and I was curious to know if her statistics training had allowed her to come to terms with this future or if the slight chance that she could see her disease recur had kept her in a near-constant state of anxiety, as it had for me.

"Humans are just really bad at processing low probabilities," Long told me as we sat at her dining table drinking iced tea. In some scenarios, people dramatically overestimate the chances that an unlikely outcome will occur. Americans spend more than seventy billion dollars a year on lottery tickets, even though the statistical probability that a particular individual will win is very low. In other cases, people underestimate the likelihood of a rare event occurring. Take natural disasters, for example. In academic studies after major floods, researchers found that residents' perceptions of the dangers they'd faced from the events often did not correspond with their actual risks.

Long told me that human beings have a lot of inherent biases when it comes to interpreting numbers. Often, people do not act rationally. In fact, the irrationality is itself predictable. Amos Tversky and Daniel Kahneman, two famous Israeli psychologists, made such biases their life's work. After our interview, Long sent me a paper the pair had published in 1981. In a study, three hundred college students were divided into two groups and asked to consider a hypothetical scenario in which a terrible disease was expected to kill six hundred people.

One group of students was told that if a particular program was adopted, two hundred people would be saved. A second group of students was told that if that particular program was adopted, four hundred people would die.

More students expressed a preference for the program in which two hundred lives would be saved than the one in which four hundred lives would be lost, even though the two outcomes were the same. "Loss just feels more painful than the equivalent gain," Long said. "And cancer is just loss after loss." How a person experiences life after cancer, Long told me, may be based not just on physical well-being but also on perception, on whether the former patient feels the pain of diagnosis, treatment and aftermath more acutely than she feels the exhilaration of having made it out alive.

Long said her cancer experience had changed her in many ways. She had lost two breasts and her ovaries, but she had a baby on the way. Her sense of femininity had been beaten down, but she was doing the most feminine thing in the world by becoming a mother.

Before her cancer, Long, ever the statistician, had thought of her life as a series of inputs and outputs. "My entire life was very much an example of investing in production and my own human capital," Long said. She earned her doctorate from Stanford and had been a professor at Yale before joining the faculty at UCLA. She'd been highly focused on her career and wanted to put off having children, even though her husband, who was five years older, was eager to start a family. "I was so invested in my career and wanting to be the best version of myself professionally. I was so focused on that and so driven by that. When I got cancer, it was just like—what's the point?" Before her diagnosis, Long was the kind of faculty member who could always be counted on to join a committee or meet with students at all hours. Since cancer, Long said she has become better at saying no. Her perception of the value of time has shifted. She told me she had recently been asked to host a weekend university admissions event for incoming female students. "That's something I really care about," Long said. "But that's a Saturday I'll never get back."

Long put her academic research skills to work after her diagnosis, making choices on her care based on what she found in the data of published studies. "But there's a set of decisions in which you can't always break it down into numbers," she said. "Like, if I knew my cancer was going to come back, would I want to have a child? I can't write down a functional form of that."

As I write this, it has been three years and one month since I was diagnosed with invasive breast cancer. Evie is now a second-grader. She is learning to play the guitar and loves to read. Two years ago, Collin and I bought our first house. It's on a large lot in Los Angeles and we are planting a small orchard, including two avocado trees, in our backyard and building a deck. We are planning for a future in which I will be alive. Every day that passes, I think less about dying.

A few years ago, I asked Sara Hurvitz, my oncologist, what would happened if my cancer recurred. She listed the drugs available to treat metastatic HER2-positive breast cancer, including Herceptin, Perjeta, Kadcyla and several others. Nearly all had been approved by the FDA in the past decade. Knowing which treatments I would get if I found myself a breast cancer patient again has helped me feel less panicked about the possibility. When a wave of anxiety about recurrence washes over me, I think about the drugs Hurvitz listed, the new medicines scientists are developing in labs across the country and the ones already in front of FDA regulators like Tatiana Prowell. In the first half of 2019 alone, two more treatment options became available to women with HER2-positive breast cancer. Herceptin can now be administered through a simple shot. And a Genentech-sponsored clinical trial showed that patients who have residual breast cancer after chemotherapy and Herceptin treatment can see their chances of recurrence cut in half if they are given Kadcyla.

Progress in the fight against breast cancer in America has been, by any measure, remarkable. The patients who have demanded better treatment, the doctors developing new drugs, and public and private investments in the cause have made this so. I am optimistic that this progress will continue. In the future, treatments that today seem as logical as the radical mastectomy once did will be discarded in favor of better ones. Breast cancer advocacy, already undergoing a cultural shift, will evolve toward a messaging strategy more appropriate for the twenty-first century. Drug treatments will become more targeted and their side effects will be more tolerable. Screening will get better. The breast cancer mortality rate will continue to fall. Women will be less afraid.

If I have learned anything from writing this book, it's that for progress to continue, we must be open to change.

acknowledgments

I was able to report and write this book only because of the efforts and openness of many others. First and foremost, thank you to the medical team who kept me alive and made it possible for me to undertake this project. This begins with Dr. Sara Hurvitz, my medical oncologist and director of the breast oncology program at the University of California, Los Angeles. It is a hard thing to put your life into someone else's hands, but Sara's confidence in the medical treatments available to me and her clearheaded way of explaining them boosted my optimism early. When you're diagnosed with a complicated disease for which treatments are changing all the time, it's a gift to have a doctor in your corner who is contributing to the global understanding of the disease. In addition to providing what I believe was the best and most modern care available anywhere in the world, Sara hugged me when I needed it and answered countless e-mails during my treatment and the writing of this book, kindnesses for which I will be forever grateful. Dr. Amy Kusske, my surgical oncologist at UCLA, cut open my chest and removed the flesh that had once been riddled with cancer. I am grateful to Dr. Kusske for her skill as a surgeon but also for her honest and patient bedside manner; she helped me come to terms with the idea of having my thirty-five-year-old breasts removed.

Thank you also to Dr. Ruth Williamson of Huntington Hospital, who designed and oversaw my radiation therapy and took extremely good care of me during that phase of my treatment.

When Collin and I met Dr. Andrew Da Lio for the first time, we felt sure he was the right person to put my body back together. He has been the chief of plastic surgery at UCLA for more than a decade and is a talented and creative technician in the operating room. But beyond that, he has a generosity of spirit that seems boundless. Early on, he gave me his cell phone number, told me to call or text anytime and to please call him Andy. Over the past few years, we have talked about so much more than cancer and breasts. Andy assisted tremendously in the reporting of this book, talking with me for hours in person and over the phone and arranging for me to observe a bilateral mastectomy and reconstruction surgery at UCLA.

The nurses, imaging technicians and other staff at UCLA treated me with extreme kindness and care when I was most vulnerable, and I will always be grateful for their skill and compassion.

I also owe an enormous debt of gratitude to the breast cancer patients who volunteered for clinical trials that informed my treatment, particularly the women who took the risk of participating in the early Herceptin trials at UCLA and elsewhere.

If I had any fortitude when it came to facing my breast cancer diagnosis and treatment, that is because of my parents. They raised my brother and me to be tough and honest, always reminding us to have perspective. Thank you for that, Mom and Dad. My parents also were among those who took care of me when I was enduring chemotherapy and then surgery. They hosted me at a rental house in Ojai, California, when I needed to rest after my first chemotherapy session and drove me to physical therapy appointments after my surgery. Thank you, Mom, for all the hugs and the meals and the gentle way you took care of me when I was sick. Thank you, Dad, for assuring me I could say anything to you about cancer and that we would talk it through.

Thanks also to my brother, Matt. There is no one else on Earth with whom I could have more fun sitting around a campfire and talking all night. Matt moved in with us during the summer of 2015 and kept me company, which was everything. The following summer, shortly after I finished the last of my treatments, he and I drove from our parents' farm in upstate New York to Los Angeles. As we made our way across Utah, Matt rode his skateboard along the Colorado River while I followed behind in my car. Thanks for doing that, Matt. You always show me new joys in life. Thanks also to the members of my extended family who supported me through my treatment, particularly my aunts Dian and Leean and my uncle Dick. In the winter of 2018, my mother-in-law, Judy Campbell, lent me her cabin in the High Sierra so I could make progress on the manuscript. My father-in-law, Bill, stocked it with firewood. Thanks to both of them for their love and generosity in this project and in life.

My chosen family on the West Coast helped me to keep moving forward during my treatment and made me laugh when I needed it most. Many friends also listened to me talk through the ideas in this book and offered suggestions that made it better. Thanks especially to Sitara Nieves, Claire Martin, Rebecca Lehrer, Cory Null, Sarah Rosenthal, Emily Dalton Smith, Megan Miller, Ky Henderson, Brandon Randolph, Marcelo

Romero, Jodie Lustgarten, Abrahm Lustgarten, and Jennifer Pauling.

As a patient, I had a front-row seat to many of the latest treatments for breast cancer and experienced them firsthand. But my personal experience was just that. Writing about what it is like to be a breast cancer patient in the modern era would not have been possible without the brave women who trusted me with their stories, including Eleonora Ford, Sheri Weitz, Nancy Lankford, Lianne Zhang, Elisa Long, Carolyn Streed and Lianne Kraemer. Others' names do not appear in the text, but their insights and wisdom informed my understanding in more ways than I can count and I am so grateful. Thank you also to the countless doctors, researchers and advocates who took time away from their important work to talk with me.

My agent, Richard Pine of Inkwell Management, was enthusiastic about this project from the very beginning and has been a terrific support throughout the process. Tracy Behar, my editor at Little, Brown, was a wise and meticulous presence. Her suggestions and edits improved the manuscript immensely. Thank you to everyone else at Inkwell and Little, Brown who assisted in bringing this book into the world.

Thank you to David Bjerklie, whose talents extend far beyond fact checking and who made this a better book. Thank you to Rachel Haik and Kaya McMullen for early research and transcription help, to Amanda Koenigsberg for citation assistance, and to Brandon Moynihan for my author photo.

Beginning a job as a university professor just as I was emerging from treatment was a rebirth. Thank you to my students and colleagues at Loyola Marymount University for giving me a new purpose. And thanks especially to journalism program director Evelyn McDonnell, English Department chair Barbara Rico and Bellarmine College of Liberal Arts dean Robbin Crabtree for supporting this project. Thanks also to the editors and writers at *New York* magazine and *Time* who helped me learn what makes a good story and to the authors whose books I often turned to when I was researching breast cancer, in particular Barron Lerner (*The Breast Cancer Wars*), Siddhartha Mukherjee (*The Emperor of All Maladies*) and Robert Bazell (*Her-2*).

In addition to always believing in me as a journalist, my husband, Collin Campbell, is a true partner in every way. During my breast cancer experience, Collin never flinched. He was a rock that I could lean against, and lean I did. I know it's a cliché to thank a spouse for the childcare he provided during the writing of a book, but I must. Thank you, Collin, for all the meals you and Evie ate alone, for all the kids' birthday parties and soccer games you attended without me so I could have a quiet house in which to write. Thank you most of all for making this book feel more like a family project than an ambition that was mine alone. And thank you to Evie for always reminding me what it is all for.

bibliography

Introduction

American Cancer Society. "Current Grants by Cancer Type." American Cancer Society. cancer.org/research/currently-funded-cancer-research/grants-by-cancer-type.html. Accessed March 2019.

Mulcahy, Nick. "The Mystery of a Common Breast Cancer Statistic." Medscape.com, August 18, 2015. medscape.com/viewarticle/849644.

Narod, Steven A., et al. "Why Have Breast Cancer Mortality Rates Declined?" *Journal of Career Policy* 5, September 2015: 8–17.

National Cancer Institute. "Cancer Stat Facts: Female Breast Cancer." Surveillance, Epidemiology, and End Results Program (SEER). seer.cancer.gov/statfacts/html/breast.html. Accessed November 2017.

National Institutes of Health. "Estimates of Funding for Various Research, Condition, and Disease Categories." National Institutes of Health, May 18, 2018. report.nih.gov/categorical_spending.aspx.

O'Shaughnessy, Joyce. "Extending Survival with Chemotherapy in Metastatic Breast Cancer." *Oncologist* 10, supp. 3 (2005): 20–29.

Sontag, Susan. "Illness as Metaphor." *New York Review of Books,* January 26, 1978.

Chapter 1. Seek and Ye Shall Find

American Cancer Society. "Breast Cancer Facts and Figures 2017–2018." American Cancer Society. cancer.org/content/dam/cancer-org/

research/cancer-facts-and-statistics/breast-cancer-facts-and-figures/breast-cancer-facts-and-figures-2017-2018.pdf.

———. "Cancer Facts and Figures 2015, Special Section: Breast Carcinoma In Situ." American Cancer Society. cancer.org/research/cancer-facts-statistics/all-cancer-facts-figures/cancer-facts-figures-2015.html.

Andersson, I., et al. "Mammographic Screening and Mortality from Breast Cancer: The Malmö Mammographic Screening Trial." *BMJ* 297, no. 6654 (1988): 943–48.

ASCO Post editors. "$13.4 Million Awarded to Study Treatment for Low-Grade Ductal Carcinoma in Situ in a Prospective, Randomized Trial." *ASCO Post,* February 25, 2016.

Autier, P., et al. "Mammography Screening and Breast Cancer Mortality in Sweden." *Journal of the National Cancer Institute* 104, no. 14 (2012): 1080–93.

Berry, Donald A., et al. "Effect of Screening and Adjuvant Therapy on Mortality from Breast Cancer." *New England Journal of Medicine* 353, no. 17 (2005): 1784–92.

Biller-Andorno, Nikola, and Peter Jüni. "Abolishing Mammography Screening Programs? A View from the Swiss Medical Board." *New England Journal of Medicine* 370, no. 21 (2014): 1965–67.

Bjurstam, Nils, et al. "The Gothenburg Breast Cancer Screening Trial: Preliminary Results on Breast Cancer Mortality for Women Aged 39–49." *Journal of the National Cancer Institute Monographs* 1997, no. 22 (1997): 53–55.

Bond, M., et al. "Systematic Review of the Psychological Consequences of False-Positive Screening Mammograms." *Health Technology Assessment* 17, no. 13 (2013): 1–170.

Broders, A. C. "Carcinoma in Situ Contrasted with Benign Penetrating Epithelium." *JAMA* 99, no. 20 (1932): 1670–74.

"Congresswoman Wasserman Schultz on Fighting Breast Cancer." *Situation Room.* CNN, March 26, 2009.

Cunningham, M. P. "The Breast Cancer Detection Demonstration Project

25 Years Later." *CA: A Cancer Journal for Clinicians* 47, no. 3 (1997): 131–33.

Esserman, Laura, et al. "Rethinking Screening for Breast Cancer and Prostate Cancer." *JAMA* 302, no. 15 (2009): 1685–92.

Federal News Service. "Hearing of the Health Subcommittee of the House Energy and Commerce Committee; Subject: Breast Cancer Screening Recommendations." December 2, 2009.

Food and Drug Administration. "MQSA National Statistics." Mammography Quality Standards Act and Program. fda.gov/radiation-emittingproducts/mammographyqualitystandardsactandprogram/facilityscorecard/ucm113858.htm. Accessed January 2018.

Frisell, J., et al. "Randomized Study of Mammography Screening—Preliminary Report on Mortality in the Stockholm Trial." *Breast Cancer Research and Treatment* 18, no. 1 (1991): 49–56.

Gøtzsche, Peter C., and Karsten Juhl Jørgensen. "Screening for Breast Cancer with Mammography." *Cochrane Database of Systematic Reviews* 6 (2013).

Gøtzsche, Peter C., and Ole Olsen. "Is Screening for Breast Cancer with Mammography Justifiable?" *Lancet* 355, no. 9198 (2000): 129–34.

Henry J. Kaiser Family Foundation. "Coverage of Breast Cancer Screening and Prevention Services." May 2018 Fact Sheet.

———. "Number of Heart Disease Deaths per 100,000 Population by Gender." State Health Facts, 2017.

Keating, Nancy L., et al. "Breast Cancer Screening in 2018: Time for Shared Decision Making." *JAMA* 319, no. 17 (2018): 1814–15.

Lerner, Barron H. *The Breast Cancer Wars.* New York: Oxford University Press, 2001.

Miller, Anthony B., et al. "Canadian National Breast Screening Study: 1. Breast Cancer Detection and Death Rates Among Women Aged 40 to 49 Years." *Journal of the Canadian Medical Association* 147, no. 10 (1992): 1459–76.

Miller, Anthony B., et al. "Canadian National Breast Screening Study: 2. Breast Cancer Detection and Death Rates Among Women Aged 50 to 59

Years." *Journal of the Canadian Medical Association* 147, no. 10 (1992): 1477–88.

Miller, Anthony B., et al. "Twenty-Five-Year Follow-Up for Breast Cancer Incidence and Mortality of the Canadian National Breast Screening Study: Randomised Screening Trial." *BMJ* 348, no. 7945 (2014): 12.

National Cancer Institute. "Cancer Stat Facts: Female Breast Cancer." Surveillance, Epidemiology, and End Results Program (SEER). seer.cancer.gov/statfacts/html/breast.html. Accessed November 2017.

———. "Cancer Stat Facts: Cervical Cancer." Surveillance, Epidemiology, and End Results Program (SEER). Accessed November 2017.

———. "Breast Cancer Risk in American Women." https://www.cancer.gov/types/breast/risk-fact-sheet.

National Institutes of Health Consensus Development Conference Statement. "Breast Cancer Screening for Women Ages 40–49." *Journal of the National Cancer Institute* 89, no. 14 (1997): 960–65.

National Institutes of Health. "Cervical Cancer." National Institutes of Health. report.nih.gov/nihfactsheets/viewfactsheet.aspx?csid=76.

National Institutes of Health Consensus Statement online. "Breast Cancer Screening." (1977) September 14–16: 1:5–8.

"New Mammogram Advice Raises Concerns." *Situation Room*. CNN, November 17, 2009.

Nyström, Lennarth, et al. "Breast Cancer Screening with Mammography: Overview of Swedish Randomised Trials." *Lancet* 341, no. 8851 (1993): 973–78.

Ong, Mei-Sing, and Kenneth D. Mandi. "National Expenditure for False-Positive Mammograms and Breast Cancer Overdiagnoses Estimated at $4 Billion a Year." *Health Affairs* 34, no. 4 (2015): 576–83.

Radhakrishnan, Archana, et al. "Physician Breast Cancer Screening Recommendations Following Guideline Changes: Results of a National Survey." *JAMA Internal Medicine* 177, no. 6 (2017): 877–78.

Sack, Kevin, and Gina Kolata. "Screening Policy Won't Change, U.S. Officials Say." *New York Times*, November 18, 2009.

"Sandra Lee Hopes Her Story Will Inspire Women to Get Mammograms." *Good Morning America.* ABC News. abcnews.go.com/GMA/video/sandra-lee-hopes-story-inspire-women-mammograms-30978492. Accessed May 2017.

"Sandra Lee Says She's Cancer Free, Discusses Her Journey to Recovery." *Good Morning America.* ABC News. abcnews.go.com/GMA/video/sandra-lee-shes-cancer-free-discusses-journey-recovery-33942104. Accessed May 2017.

Surveillance, Epidemiology, and End Results (SEER) Program. DevCan database: "SEER 21 Incidence and Mortality, 2000–2016, with Kaposi Sarcoma and Mesothelioma." National Cancer Institute, DCCPS, Surveillance Research Program, Cancer Statistics Branch, released April 2019. Underlying mortality data provided by NCHS. www.cdc.gov/nchs.

Shapiro, Sam, et al. "Periodic Breast Cancer Screening in Reducing Mortality from Breast Cancer." *JAMA* 215, no. 11 (1971): 1777–85.

Shapiro, Sam, et al. "Ten- to Fourteen-Year Effect of Screening on Breast Cancer Mortality." *Journal of the National Cancer Institute* 69, no. 2 (1982): 349–55.

Shapiro, Sam, et al. "Selection, Follow-Up, and Analysis in the Health Insurance Plan Study: A Randomized Trial with Breast Cancer Screening." *National Cancer Institute Monograph* 67 (1985): 65–74.

Shapiro, Sam. "Periodic Screening for Breast Cancer: The HIP Randomized Controlled Trial." *Journal of the National Cancer Institute Monographs* 1997, no. 22 (1997): 27–30.

Tan, Siang Yong, and Yvonne Tatsumura. "George Papanicolaou (1883–1962): Discoverer of the Pap Smear." *Singapore Medical Journal* 56, no. 10 (2015): 586–87.

Tosteson, Anna N. A., et al. "Consequences of False-Positive Screening Mammograms." *JAMA Internal Medicine* 174, no. 6 (2014): 954–61.

U.S. Preventive Services Task Force. "Breast Cancer: Screening." January 2016. uspreventiveservicestaskforce.org/Page/Document/UpdateSummaryFinal/breast-cancer-screening1.

———. "Screening for Breast Cancer: U.S. Preventive Services Task Force Recommendation Statement." *Annals of Internal Medicine* 151, no. 10 (2009): 716–26.

Van Cleef, A., et al. "Current View on Ductal Carcinoma In Situ and Importance of the Margin Thresholds: A Review." *Facts, Views and Vision: Issues in Obstetrics, Gynaecology and Reproductive Health* 10, no. 2 (2018): 210–18.

Ward, Elizabeth M., et al. "Cancer Statistics: Breast Cancer In Situ." *CA: A Cancer Journal for Clinicians* 65, no. 6 (2015): 481–95.

Welch, H. Gilbert, et al. "Trends in Metastatic Breast and Prostate Cancer—Lessons in Cancer Dynamics." *New England Journal of Medicine* 373, no. 18 (2015): 1685–87.

Chapter 2. CSI: Breasts

Begley, Sharon. "Grail's Cancer Blood Test Shows 'Proof of Principle,' But Challenges Remain." *Stat,* June 2, 2018. statnews.com/2018/06/02/grail-cancer-blood-test-asco/.

Cho, Nariya, et al. "Breast Cancer Screening with Mammography Plus Ultrasonography or Magnetic Resonance Imaging in Women 50 Years or Younger at Diagnosis and Treated with Breast Conservation Therapy." *JAMA Oncology* 3, no. 11 (2017): 1495–1502.

Comparison of Molecular Breast Imaging and Digital Breast Tomosynthesis for Screening in Women with Dense Breasts (MBI-DBT). ClinicalTrials.gov. Last updated August 16, 2018. clinicaltrials.gov/ct2/show/NCT03220893?term=mbi%2C+deborah+rhodes&cond=breast+cancer&rank=2.

Digital Tomosynthesis Mammography and Digital Mammography in Screening Patients for Breast Cancer. ClinicalTrials.gov. Last updated December 14, 2018. clinicaltrials.gov/ct2/show/NCT03233191?term=tmist&rank=1.

Drukker, C. A., et al. "Mammographic Screening Detects Low-Risk Tumor Biology Breast Cancers." *Breast Cancer Research and Treatment* 144, no. 1 (2014): 103–11.

ECOG-ACRIN Cancer Research Group. TMIST Breast Cancer Screening Trial. ecog-acrin.org/tmist. Accessed March 2018.

Eddy, David M. "The Frequency of Cervical Cancer Screening. Comparison of a Mathematical Model with Empirical Data." *Cancer* 60, no. 5 (1987): 1117–22.

Esserman, Laura, et al. "Overdiagnosis and Overtreatment in Cancer: An Opportunity for Improvement." *JAMA* 310, no. 8 (2013): 797–98.

Esserman, Laura, and Christina Yau. "Rethinking the Standard for Ductal Carcinoma In Situ Treatment." *JAMA Oncology* 1, no. 7 (2015): 881–83.

Esserman, Laura, and Ian Thompson. "Solving the Overdiagnosis Dilemma." *Journal of the National Cancer Institute* 102, no. 9 (2010): 582–83.

Esserman, Laura, et al. "The WISDOM Study: Breaking the Deadlock in the Breast Cancer Screening Debate." *Nature Partner Journals: Breast Cancer* 3, no. 34 (2017).

Fernandez, Elizabeth. "UCSF to Study Benefits of Personal Approach to Breast Cancer Screening." UCSF News Center, March 25, 2015. ucsf.edu/news/2015/03/124241/ucsf-gets-major-contract-study-benefits-personal-approach-breast-cancer.

Food and Drug Administration. "FDA Advances Landmark Policy Changes to Modernize Mammography Services and Improve Their Quality." U.S. Food and Drug Administration, March 27, 2019. https://www.fda.gov/NewsEvents/Newsroom/PressAnnouncements/ucm634509.htm.

Grady, Denise. "Nancy Cappello, Breast Cancer Activist, Is Dead at 66." *New York Times*, November 28, 2018.

Hafner, Katie. "A Strong Second Opinion." *New York Times*, September 28, 2015.

Hruska, Carrie B. "Molecular Breast Imaging for Screening in Dense Breasts: State of the Art and Future Directions." *American Journal of Roentgenology* 208, no. 2 (2017): 275–83.

Hruska, Carrie B., et al. "Diagnostic Workup and Costs of a Single Supplemental Molecular Breast Imaging Screen of Mammographically Dense Breasts." *American Journal of Roentgenology* 204, no. 6 (2015): 1345–53.

Johnston, Brian S., et al. "Can Preoperative Breast MRI Help Predict DCIS Recurrence?" *Journal of Clinical Oncology* 34, no. 15 (2016).

Kuhl, Christiane, et al. "Abbreviated Breast Magnetic Resonance Imaging (MRI): First Postcontrast Subtracted Images and Maximum-Intensity Projection—A Novel Approach to Breast Cancer Screening with MRI." *Journal of Clinical Oncology* 32, no. 22 (2014): 2304–10.

Kuhl, Christiane K. "The Changing World of Breast Cancer: A Radiologist's Perspective." *Investigative Radiology* 50, no. 9 (2015): 615–28.

Kuhl, Christiane, et al. "Breast MRI Screening of Women at Average Risk of Breast Cancer: An Observational Cohort Study." *Journal of Clinical Oncology* 33, no. 28 (2015): 1.

———. "Supplemental Breast MR Imaging Screening of Women with Average Risk of Breast Cancer." *Radiology* 283, no. 2 (2017): 361–70.

Morris, Elizabeth. "Rethinking Breast Cancer Screening: Ultra FAST Breast Magnetic Resonance Imaging." *Journal of Clinical Oncology* 32, no. 22 (2014): 2281–83.

O'Donoghue, Cristina, et al. "Aggregate Cost of Mammography Screening in the United States: Comparison of Current Practice and Advocated Guidelines." *Annals of Internal Medicine* 160, no. 3 (2014): 145–53.

Pisano, Etta, et al. "Diagnostic Performance of Digital Versus Film Mammography for Breast Cancer Screening." *New England Journal of Medicine* 353, no. 17 (2005): 1773–83.

———. "Patient Compliance in Mobile Screening Mammography." *Academic Radiology* 2, no. 12 (1995): 1067–72.

Rhodes, Deborah J., et al. "Dedicated Dual-Head Gamma Imaging for Breast Cancer Screening in Women with Mammographically Dense Breasts." *Radiology* 258 (2011): 106–18.

———. "Molecular Breast Imaging at Reduced Radiation Dose for Supplemental Screening in Mammographically Dense Breasts." *American Journal of Roentgenology* 204, no. 2 (2015): 241–51.

Chapter 3. Diagnosis

99% Invisible. Episode 173, "How a Simple Ribbon Took AIDS from Taboo to Trendy Cause." July 21, 2015. 99percentinvisible.org/episode/awareness/.

American Society of Breast Surgeons. "Consensus Guideline on Genetic Testing for Hereditary Breast Cancer." breastsurgeons.org/docs/statements/Consensus-Guideline-on-Genetic-Testing-for-Hereditary-Breast-Cancer.pdf. Accessed December 2018.

Belman, Orli. "Hormones and Breast Cancer." PBS.org. June 1998. pbs.org/wgbh/pages/frontline/shows/nature/disrupt/breast.html.

Chillisauce (blog). "The Biggest Bra on the Planet." chillisauce.com/blog/post-63101c7489450b8a709c.

Department of Health and Human Services. *Report to Congress: The Long Island Breast Cancer Study Project*. November 2004. epi.grants.cancer.gov/past-initiatives/LIBCSP/ReptoCong_508compliant.pdf.

Desaid, Sunita, and Anupam B. Jena. "Do Celebrity Endorsements Matter? Observational Study of *BRCA* Gene Testing and Mastectomy Rates After Angelina Jolie's *New York Times* Editorial." *BMJ* 355, no. 8086 (December 2016): 491–93.

Ehrenreich, Barbara. "Welcome to Cancerland: A Mammogram Leads to a Cult of Pink Kitsch." *Harper's*, November 2001.

Fagin, Dan. "Tattered Hopes: A $30-Million Federal Study of Breast Cancer and Pollution on LI Has Disappointed Activists and Scientists Alike." *Newsday*, July 28, 2002.

———. "Tattered Hopes—Still Searching: A Computer Mapping System Was Supposed to Help Unearth Information About Breast Cancer and the Environment." *Newsday*, July 30, 2002.

———. "Tattered Hopes—Study in Frustration: Ambitious Search for Links Between Pollution and Breast Cancer on LI Was Hobbled from the Start, Critics Say." *Newsday*, July 29, 2002.

Farvid, M. S., et al. "Consumption of Red and Processed Meat and Breast Cancer Incidence: A Systematic Review and Meta-Analysis of Prospective Studies." *International Journal of Cancer* 143, no. 11 (December 2018): 2787–99.

Fischler, Marcelle S. "Long Island Journal: After Breast Cancer, Changing the World One House at a Time." *New York Times,* January 8, 2006.

Helwick, Caroline. "Dr. Mary-Claire King Proposes Population Screening in All Young Women for *BRCA* Mutations." *ASCO Post,* February 10, 2015.

Hendrix, Kathleen. "Peach Corps: Activism: Breast Cancer Has Afflicted Her Grandmother, Sister and Daughter, So Charlotte Haley Is Urging People to Wear Ribbons to 'Wake Up' America." *Los Angeles Times,* August 20, 1992.

Interagency Breast Cancer and Environmental Research Coordinating Committee. "Breast Cancer and the Environment: Prioritizing Prevention." February 2013.

King, Mary-Claire. "'The Race' to Clone *BRCA1.*" *Science* 343, no. 6178 (March 28, 2014): 1462–65.

Kolata, Gina. "Epidemic That Wasn't." *New York Times,* August 29, 2002.

———. "L.I. Cancer Found to Be Explainable." *New York Times,* December 19, 1992.

Liptak, Adam. "Justices, 9–0, Bar Patenting Human Genes." *New York Times,* June 13, 2013.

Miller, Elizabeth Kiggen. "Proposals Awaited for L.I. Breast Cancer Study." *New York Times,* August 9, 1998.

Mnookin, Seth. *The Panic Virus: The True Story Behind the Vaccine-Autism Controversy.* New York: Simon and Schuster, 2011.

National Breast Cancer Coalition. "Legislative Accomplishments." breastcancerdeadline2020.org/get-involved/public-policy.

Nelson, N. J. "Migrant Studies Aid the Search for Factors Linked to Breast Cancer Risk." *Journal of the National Cancer Institute* 98, no. 7 (April 2006): 436–38.

Olive Oil for High Risk Breast Cancer Prevention in Women. ClinicalTrials.gov. Last updated July 24, 2018. clinicaltrials.gov/ct2/show/NCT02068092?term=NCT02068092&rank=1.

Peterson, Pia. "A Cover Photo's Legacy, 25 Years Later." *New York Times,* August 14, 2018.

Pickert, Kate. "Lessons from Angelina: The Tricky Calculus of Cancer Testing." Time.com, May 15, 2013. nation.time.com/2013/05/15/lessons-from-angelina-the-tricky-calculus-of-cancer-testing/.

Purdum, Todd S. "Virginia Clinton Kelley, 70, President's Mother, Is Dead." *New York Times,* January 7, 1994.

Quinn, Audrey. "Before Pink Became Synonymous with Breast Cancer, There Was Peach." *Pulse,* July 23, 2015. whyy.org/segments/before-pink-became-synonymous-with-breast-cancer-there-was-peach/.

Schemo, Diana Jean. "L.I. Breast Cancer Is Possibly Linked to Chemical Sites." *New York Times,* April 13, 1994.

Tomasetti, Cristian, and Bert Vogelstein. "Variation in Cancer Risk Among Tissues Can Be Explained by the Number of Stem Cell Divisions." *Science* 347, no. 6217 (2015): 78–81.

Tomasetti, Cristian, et al. "Stem Cell Divisions, Somatic Mutations, Cancer Etiology, and Cancer Prevention." *Science* 355, no. 6331 (2017): 1330–34.

Chapter 4. Pink Vibes

Bassett, Laura. "Karen Handel, Susan. G. Komen's Anti-Abortion VP, Drove Decision to Defund Planned Parenthood." *Huffington Post,* February 5, 2012. huffingtonpost.com/2012/02/05/karen-handel-susan-g-komen-decision-defund-planned-parenthood_n_1255948.html.

Black, Shirley Temple. "Don't Sit Home and Be Afraid." *McCall's* (February 1973): 82, 114–16.

Brinker, Nancy G. *Promise Me: How a Sister's Love Launched the Global Movement to End Breast Cancer.* New York: Crown Archetype, 2010.

Handel, Karen. *Planned Bullyhood: The Truth About the Planned Parenthood Funding Battle with Susan G. Komen for the Cure.* New York: Howard Books, 2012.

Holland, Jimmie C., and Sheldon Lewis. *The Human Side of Cancer: Living with Hope, Coping with Uncertainty.* New York: HarperCollins, 1999.

Ing, Roy, et al. "Unilateral Breast-Feeding and Breast Cancer." *Lancet* 310, no. 8029 (1977): 124–27.

Karen Handel for Governor. "Karen Handel on Life and Planned Parenthood." web.archive.org/web/20100718101249/http://blog.karenhandel.com/2010/07/karen-handel-on-life-and-planned-parenthood.

Khan, Huma. "Susan G. Komen Apologizes for Cutting Off Planned Parenthood Funding." *Washington Post*, February 3, 2012.

Kinsman, Kat. "Activists Call Foul on KFC Bucket Campaign." CNN.com, April 28, 2010. cnn.com/2010/LIVING/homestyle/04/28/kfc.pink.bucket.campaign/index.html.

Kushner, Rose. *Breast Cancer: A Personal History and an Investigative Report*. New York: Harcourt Brace Jovanovich, 1975.

Lerner, Barron H. *The Breast Cancer Wars*. New York: Oxford University Press, 2001.

Love, Susan. *Dr. Susan Love's Breast Book*. Boston: Da Capo, 2015.

Miller, Sunlen. "Senators Urge Komen Foundation to Reverse 'Troubling' Planned Parenthood Funding Cut." ABC News, February 3, 2012.

Morgan, David. "Catholic Bishops Pressured Komen Over Planned Parenthood." Reuters, March 15, 2012. reuters.com/article/us-usa-komen-catholic/catholic-bishops-pressured-komen-over-planned-parenthood-idUSBRE82E12Q20120315.

Mullany, Anjali. "Can American Apparel's CEO Mend Its Seams?" *Fast Company*, July 22, 2016. fastcompany.com/3061612/can-american-apparels-ceo-mend-its-seams.

Non-Profit Times. "*NPT* Top 100: An In-Depth Study of America's Largest Nonprofits." 2012 and 2014.

Preston, Jennifer. "After Outcry, a Senior Official Resigns at Komen." *New York Times*, February 7, 2012.

Rollin, Betty. *First, You Cry*. New York: J. B. Lippincott, 1976.

Sun, Lena H. "Komen Names Judith Salerno as CEO to Replace Nancy Brinker." *Washington Post*, June 17, 2013.

Susan G. Komen Breast Cancer Foundation. *2011–2012 Annual Report*. Dallas: Susan G. Komen Breast Cancer Foundation, 2012.

———. *Fiscal Year 2013 Annual Report*. Dallas: Susan G. Komen Breast Cancer Foundation, 2013.

———. "Susan G. Komen Sets Bold Goal to Reduce U.S. Breast Cancer Deaths by 50% in 10 years," September 13, 2016. komennewengland .org/BIGBOLDGOAL.

United States. Department of the Treasury. Internal Revenue Service. "Form 990." Susan G. Komen Breast Cancer Foundation, Group. 2011.

United States. Department of the Treasury. Internal Revenue Service. "Form 990." Susan G. Komen Breast Cancer Foundation, Inc. 2011.

United State. Department of the Treasury. Internal Revenue Service. "Form 990." Susan G. Komen Breast Cancer Foundation, Group. 2017.

United States. Department of the Treasury. Internal Revenue Service. "Form 990." Susan G. Komen Breast Cancer Foundation, Inc. 2017.

Chapter 5. Lady Parts

Abderrahman, Balkees, et al. "Rethinking Extended Adjuvant Antiestrogen Therapy to Increase Survivorship in Breast Cancer." *JAMA Oncology* 4, no. 1 (2018): 15–16.

AstraZeneca. "National Breast Cancer Awareness Month." AstraZeneca. astrazeneca-us.com/sustainability/healthcare-foundation/national -breast-cancer-awareness-month-.html. Accessed June 2018.

Beatson, G. T. "On the Treatment of Inoperable Cases of Carcinoma of the Mamma: Suggestions for a New Method of Treatment, with Illustrative Cases." *Lancet*, no. 2 (1896): 104–7, 162–65.

Beral, Valerie, et al. "Breast Cancer Risk in Relation to the Interval Between Menopause and Starting Hormone Therapy." *Journal of the National Cancer Institute* 103, no. 4 (2011): 296–305.

Boyd, Stanley. "On Oöphorectomy in Cancer of the Breast." *BMJ* 2, no. 2077 (1900): 1161–67.

Britt, K., and R. Short. "The Plight of Nuns: Hazards of Nulliparity." *Lancet* 379, no. 9834 (2012): 2322–23.

Crew, Katherine D., et al. "Prevalence of Joint Symptoms in Postmenopausal Women Taking Aromatase Inhibitors for Early-Stage Breast Cancer." *Journal of Clinical Oncology* 25, no. 25 (2007): 3877–83.

Early Breast Cancer Trialists' Collaborative Group. "Aromatase Inhibitors Versus Tamoxifen in Early Breast Cancer: Patient-Level Meta-Analysis of the Randomised Trials." *Lancet* 386, no. 10001 (2015): 1341–52.

Easton, John. "Jensen Wins Lasker for Research on Estrogen Receptors." *University of Chicago Chronicle* 24, no. 2 (2004).

Francis, P. A., et al. "Tailoring Adjuvant Endocrine Therapy for Premenopausal Breast Cancer." *New England Journal of Medicine* 379, no. 2 (2018): 122–37.

Fraumeni, J. F., Jr., et al. "Cancer Mortality Among Nuns: Role of Marital Status in Etiology of Neoplastic Disease in Women." *Journal of the National Cancer Institute* 42, no. 3 (1969): 455–68.

Gupta, Sujata. "Profile of V. Craig Jordan." *Proceedings of the National Academy of Sciences* 108, no. 47 (November 2011). ncbi.nlm.nih.gov/pmc/articles/PMC3223455/.

Jensen, Elwood V. "From Chemical Warfare to Breast Cancer Management." *Nature Medicine* 10 (2004): 1018–21.

Jordan, V. Craig. "Tamoxifen as the First Targeted Long-Term Adjuvant Therapy for Breast Cancer." *Endocrine-Related Cancer* 21, no. 3 (2014): R235–46.

———. *Tamoxifen: A Guide for Clinicians and Patients.* New York: PRR, 1996.

———. "Tamoxifen: A Most Unlikely Pioneering Medicine." *Nature Reviews: Drug Discovery* 2, no. 3 (2003): 205–13.

———. "Tamoxifen (ICI46,474) as a Targeted Therapy to Treat and Prevent Breast Cancer." *British Journal of Pharmacology* 147 (2006): S269–76.

Kadakia, Kunal C., et al. "Patient-Reported Outcomes and Early Discontinuation in Aromatase Inhibitor–Treated Postmenopausal Women with Early Stage Breast Cancer." *Oncologist* 21, no. 5 (2016): 539–46.

Lambertini, Matteo, et al. "Safety of Pregnancy in Patients with History of Estrogen Receptor Positive Breast Cancer: Long-Term Follow-Up

Analysis from a Multicenter Study." *Journal of Clinical Oncology* 35 (18 supplement) June 3, 2017.

Lane-Claypon, J. E. "A Further Report on Cancer of the Breast with Special Reference to Its Associated Antecedent Conditions." *Reports on Public Health and Medical Subjects* 32 (1926).

Love, R. R., and J. Phillips. "Oophorectomy for Breast Cancer: History Revisited." *Journal of the National Cancer Institute* 94, no. 19 (2002): 1433–34.

MacMahon, B., et al. "Age at First Birth and Breast Cancer Risk." *Bulletin of the World Health Organization* 43, no. 2 (1970): 209–21.

Manson, JoAnn E., et al. "Menopause Hormone Therapy and Long-term All-Cause and Cause-Specific Mortality: The Women's Health Initiative Randomized Trials." *JAMA* 318, no. 10 (2017): 927–38.

Moore, David D. "A Conversation with Elwood Jensen." *Annual Review of Physiology* 74 (2012): 1–11.

Mørch, Lina S., et al. "Contemporary Hormonal Contraception and the Risk of Breast Cancer." *New England Journal of Medicine* 377, no. 23 (2017): 2228–39.

Partridge, Ann, et al. "Adherence to Initial Adjuvant Anastrozole Therapy Among Women with Early-Stage Breast Cancer." *Journal of Clinical Oncology* 26, no. 4 (2008): 556–62.

Women's Health Initiative Investigators. "Risks and Benefits of Estrogen Plus Progestin in Healthy Postmenopausal Women: Principal Results from the Women's Health Initiative Randomized Controlled Trial." *JAMA* 288, no. 3 (2002): 321–33.

Chapter 6. Not Your Mother's Chemotherapy

Altman, Lawrence K. "Insurer to Finance Test of a Treatment for Breast Cancer." *New York Times,* November 12, 1990.

Associated Press. "New Source of Cancer Drug Spares Yew Tree." *New York Times,* January 31, 1993.

———. "New Version of Taxol Is Approved by F.D.A." *New York Times,* December 13, 1994.

Brownlee, Shannon, and Dan Winters. "Bad Science and Breast Cancer." *Discover,* March 27, 1990.

Bryan, Jenny. "How Bark from the Pacific Yew Tree Improved the Treatment of Breast Cancer." *Pharmaceutical Journal*. September 21, 2011: 369.

Cauvin, Henri E. "Cancer Researcher in South Africa Who Falsified Data Is Fired." *New York Times,* March 11, 2000.

DeVita, Vincent T. "A History of Cancer Chemotherapy." *Cancer Research* 68, no. 21 (2008): 8643–53.

Eckholm, Erik. "$89 Million Awarded Family Who Sued H.M.O." *New York Times*, December 30, 1993.

———. "The Price of Hope: Medicine's Disputed Frontier." *New York Times,* September 19, 1991.

Goodman, Jordan, and Vivien Walsh. *The Story of Taxol: Nature and Politics in the Pursuit of an Anti-Cancer Drug.* New York: Cambridge University Press, 2001.

Gottlieb, Scott. "Breast Cancer Researcher Accused of Serious Scientific Misconduct." *BMJ* 320, no. 7232 (2000): 398.

Grady, Denise. "Breast Cancer Researcher Admits Falsifying Data." *New York Times,* February 5, 2000.

———. "Conference Divided Over High-Dose Breast Cancer Treatment." *New York Times,* May 18, 1999.

———. "Doubts Raised on a Breast Cancer Procedure." *New York Times,* April 16, 1999.

Holmes, F. A., et al. "The M. D. Anderson Cancer Center Experience with Taxol in Metastatic Breast Cancer." *Journal of the National Cancer Institute Monographs* 15 (1993): 161–69.

Hortobagyi, Gabriel N., et al. "Randomized Trial of High-Dose Chemotherapy and Blood Cell Autografts for High-Risk Primary Breast Carcinoma." *Journal of the National Cancer Institute* 92, no. 3 (2000): 225–33.

"Is More Better? ASCO Plenary Session Opens Debate on High-Dose Chemotherapy." *Oncologist* 4, no. 3 (1999): 269–74.

Kolata, Gina, and Kurt Eichenwald. "Hope for Sale: Business Thrives on Unproven Care, Leaving Science Behind." *New York Times,* October 3, 1999.

———. "Insurer Drops a Therapy for Breast Cancer." *New York Times,* February 16, 2000.

Kolata, Gina. "Tree Yields a Cancer Treatment, But Ecological Cost May Be High." *New York Times,* May 13, 1991.

———. "Women Resist Trials to Test Marrow Transplants." *New York Times,* February 15, 1995.

Mello, Michelle M., and Troyen A. Brennan. "The Controversy Over High-Dose Chemotherapy with Autologous Bone Marrow Transplant for Breast Cancer." *Health Affairs* 20, no. 5 (2001).

National Cancer Institute. "A Story of Discovery: Natural Compound Helps Treat Breast and Ovarian Cancers." Cancer.gov. March 31, 2015. cancer.gov/research/progress/discovery/taxol.

———. "Taxol (NSC 125973)." Cancer.gov. dtp.cancer.gov/timeline/flash/success_stories/S2_taxol.htm. Accessed July 2018.

National Historic Chemical Landmark Program. *The Discovery of Camptothecin and Taxol.* Triangle Park, NC: American Chemical Society, 2003.

Norton, Rob. "Owls, Trees, and Ovarian Cancer." *Fortune,* February 5, 1996.

O'Shaughnessy, Joyce. "High Dose Chemotherapy for Breast Cancer: Taking Stock." *Oncologist* 5, no. 1 (2000): 14–17.

Peters, William P., et al. "High-Dose Chemotherapy and Autologous Bone Marrow Support as Consolidation After Standard-Dose Adjuvant Therapy for High-Risk Primary Breast Cancer." *Journal of Clinical Oncology* 11, no. 6 (1993): 1132–43.

Pusztai, Lajos, and Gabriel N. Hortobagyi. "Discouraging News for High-Dose Chemotherapy in High-Risk Breast Cancer." *Lancet* 352, no. 9127 (1998): 501–2.

Rosenthal, Elisabeth. "Patient's Marrow Emerges as Key Cancer Tool." *New York Times,* March 27, 1990.

Runowicz, C. D. "Taxol in Ovarian Cancer." *Cancer* 15, no. 71 (1993): 1591–96.

"USDA Botanist Arthur Barclay Dies." *Washington Post,* November 16, 2003. washingtonpost.com/archive/local/2003/11/16/usda-botanist-arthur-barclay-dies/e198c855-7e1e-48d8-9720-1c7c85a3f31b/?noredirect=on&utm_term=.dac19418645f.

Weinstock, Cheryl. "Lawyers Debate the Insurability of Bone-Marrow Transplants." *New York Times,* March 20, 1994.

Weiss, Raymond B., et al. "An On-Site Audit of the South African Trial of High-Dose Chemotherapy for Metastatic Breast Cancer and Associated Publications." *Journal of Clinical Oncology* 19, no. 11 (2001): 2771–77.

Welch, H. Gilbert. "Presumed Benefit: Lessons from the American Experience with Marrow Transplantation for Breast Cancer." *BMJ* 324, no. 7345 (2002): 1088–92.

Chapter 7. Pick Your Poison

Giordano, Sharon H., et al. "Decline in the Use of Anthracyclines for Breast Cancer." *Journal of Clinical Oncology* 30, no. 18 (2012): 2232–39.

Giuliano, Armando E. "Eighth Edition of the AJCC Cancer Staging Manual: Breast Cancer." *Annals of Surgical Oncology* 25, no. 7 (2018): 1783–85.

Greene, Frederick L., and Leslie H. Sobin. "The Staging of Cancer: A Retrospective and Prospective Appraisal." *CA: A Cancer Journal for Clinicians* 58, no. 3 (2008): 180–90.

Hortobagyi, Gabriel N., et al. "New and Important Changes in the TNM Staging System for Breast Cancer." *American Society of Clinical Oncology Educational Book* 38 (2018): 457–67.

International Union Against Cancer. "TNM History, Evolution and Milestones." International Union Against Cancer. uicc.org/sites/main/files/private/History_Evolution_Milestones_0.pdf. Accessed December 2017.

Kohler, Betsy A., et al. "Annual Report to the Nation on the Status of Cancer, 1975–2011, Featuring Incidence of Breast Cancer Subtypes by Race/Ethnicity, Poverty, and State." *Journal of the National Cancer Institute* 107, no. 6 (2015): 1–25.

Loncaster, J., et al. "Impact of Oncotype DX Breast Recurrence Score Testing on Adjuvant Chemotherapy Use in Early Breast Cancer: Real World Experience in Greater Manchester, UK." *European Journal of Surgical Oncology* 43, no. 5 (2017): 931–37.

Makower, Della, and Joseph A. Sparano. "Breast Cancer Management in the TAILORx Era: Less Is More." NAM Perspectives. Discussion Paper, National Academy of Medicine, December 17, 2018.

Memorial Sloan Kettering Cancer Center. "Breast and Imaging Center Opens." Memorial Sloan Kettering Cancer Center, December 1, 2009. mskcc.org/blog/breast-and-imaging-opens.

National Cancer Institute. "Review of Staging Systems." https://training.seer.cancer.gov/collaborative/intro/systems_review.html. Accessed August 2018.

Norton, Larry. "A Gompertzian Model of Human Breast Cancer Growth." *Cancer Research* 48, no. 24 (1988): 7067–71.

Norton, Larry, et al. "Predicting the Course of Gompertzian Growth." *Nature* 264 (1976): 542–45.

Paik, Soonmyung, et al. "A Multigene Assay to Predict Recurrence of Tamoxifen-Treated, Node-Negative Breast Cancer." *New England Journal of Medicine* 351 (2004): 2817–26.

Siegel, Rebecca L., et al. "Cancer Statistics, 2018." *CA: A Cancer Journal for Clinicians* 68, no. 1 (2018): 7–30.

Slamon, Dennis, et al. "Adjuvant Trastuzumab in HER2-Positive Breast Cancer." *New England Journal of Medicine* 365 (2011): 1273–83.

Sparano, Joseph A., et al. "Adjuvant Chemotherapy Guided by a 21-Gene Expression Assay in Breast Cancer." *New England Journal of Medicine* 379 (2018): 111–21.

Chapter 8. Targets

Altman, Lawrence K. "Drug Is Shown to Shrink Tumors in Breast Cancer Characterized by Gene Defect." *New York Times,* May 18, 1998.

Bazell, Robert. *Her-2: The Making of Herceptin, a Revolutionary Treatment for Breast Cancer.* New York: Random House, 1998.

Finn, Richard S., et al. "The Cyclin-Dependent Kinase 4/6 Inhibitor Palbociclib in Combination with Letrozole Versus Letrozole Alone as First-Line Treatment of Oestrogen Receptor-Positive, HER2-Negative, Advanced Breast Cancer (PALOMA-1/TRIO-18): A Randomised Phase 2 Study." *Lancet Oncology* 16, no. 1 (2015): 25–35.

———. "Palbociclib and Letrozole in Advanced Breast Cancer." *New England Journal of Medicine* 375 (2016): 1925–36.

Fraser, Lauren. "Cloning Insulin." Genentech, April 7, 2016. gene.com/stories/cloning-insulin.

Garber, Ken "The Cancer Drug That Almost Wasn't." *Science* 345, no. 6199 (2014): 865–67.

Gianni, L., et al. "Efficacy and Safety of Neoadjuvant Pertuzumab and Trastuzumab in Women with Locally Advanced, Inflammatory, or Early HER2-Positive Breast Cancer (NeoSphere): A Randomised Multicentre, Open-Label, Phase 2 Trial." *Lancet Oncology* 13, no. 1 (2012): 25–32.

Hurvitz, Sara A., et al. "Neoadjuvant Trastuzumab, Pertuzumab, and Chemotherapy Versus Trastuzumab Emtansine plus Pertuzumab in Patients with HER2-Positive Breast Cancer (KRISTINE): A Randomised, Open-Label, Multicentre, Phase 3 Trial." *Lancet Oncology* 19, no. 1 (2018): 115–26.

———. "Pathologic Complete Response (pCR) Rates After Neoadjuvant Trastuzumab Emtansine (T-DM1 [K]) + Pertuzumab (P) Versus Docetaxel + Carboplatin + Trastuzumab + P (TCHP) Treatment in Patients with HER2-Positive (HER2+) Early Breast Cancer (EBC) (KRISTINE)." *Journal of Clinical Oncology* 34, no. 15 (2016): 500.

Memorial Sloan Kettering Cancer Center. *2017 Annual Report*. New York: Memorial Sloan Kettering Cancer Center, 2017.

The Nobel Prize in Physiology or Medicine, 1989. "J. Michael Bishop, Harold Varmus." nobelprize.org/prizes/medicine/1989/press-release.

Slamon, D. J., et al. "Human Breast Cancer: Correlation of Relapse and Survival with Amplification of the *HER-2/Neu* Oncogene." *Science* 235, no. 4785 (1987): 177–82.

———. "Studies of the *HER-2/Neu* Proto-Oncogene in Human Breast and Ovarian Cancer." *Science* 244, no. 4905 (1989): 707–12.

Timmerman, Luke. "Genentech's Souped-Up Herceptin: The Odyssey Toward a More Powerful Breast Cancer Drug." *Xconomy*, June 14, 2010.

UCLA. *2016–2017 Annual Financial Report*. Los Angeles: UCLA, 2017.

Verma, Sunil, et al. "Trastuzumab Emtansine for HER2-Positive Advanced Breast Cancer." *New England Journal of Medicine* 367 (2012): 1783–91.

von Minckwitz, Gunter, et al. "Adjuvant Pertuzumab and Trastuzumab in Early HER2-Positive Breast Cancer." *New England Journal of Medicine* 377 (2017): 122–31.

Chapter 9. From Scalpels to Sentinels

Barron, Alison U., et al. "Association of Low Nodal Positivity Rate Among Patients with ERBB2-Positive or Triple-Negative Breast Cancer and Breast Pathologic Complete Response to Neoadjuvant Chemotherapy." *JAMA Surgery* 153, no. 12 (2018): 1120–26.

Cortazar, Patricia, et al. "Pathological Complete Response and Long-Term Clinical Benefit in Breast Cancer: The CTNeoBC Pooled Analysis." *Lancet* 384, no. 9938 (2014): 164–72.

Crile, George. *What Women Should Know About the Breast Cancer Controversy*. New York: Macmillan, 1973.

Donker, M., et al. "Radiotherapy or Surgery of the Axilla After a Positive Sentinel Node in Breast Cancer (EORTC 10981-22023 AMAROS): A Randomised, Multicentre, Open-Label, Phase 3 Non-Inferiority Trial." *Lancet Oncology* 15, no. 12 (2014): 1303–10.

Eliminating Surgery After Systemic Therapy in Treating Patients with HER2 Positive or Triple Negative Breast Cancer. ClinicalTrials.gov. Last updated December 17, 2018. https://clinicaltrials.gov/ct2/show/NCT02945579.

Fagerlin, A., et al. "An Informed Decision? Breast Cancer Patients and Their Knowledge About Treatment." *Patient Education and Counseling* 64, nos. 1–3 (2006): 303–12.

Fisher, Bernard, et al. "Five-Year Results of a Randomized Clinical Trial Comparing Total Mastectomy and Segmental Mastectomy with or Without Radiation in the Treatment of Breast Cancer." *New England Journal of Medicine* 312, no. 11 (1985): 665–73.

———. "Conservative Surgery for the Management of Invasive and Noninvasive Carcinoma of the Breast: NSABP Trials." *World Journal of Surgery* 18, no. 1 (1994): 63–69.

———. "Findings from NSABP Protocol No. B-04: Comparison of Radical Mastectomy with Alternative Treatments. II. The Clinical and Biologic Significance of Medial-Central Breast Cancers." *Cancer* 48, no. 8 (1981): 1863–72.

———. "Twenty-Year Follow-Up of a Randomized Trial Comparing Total Mastectomy, Lumpectomy, and Lumpectomy Plus Irradiation for the Treatment of Invasive Breast Cancer." *New England Journal of Medicine* 347, no. 16 (2002): 1233–41.

Giuliano, A. E., et al. "Axillary Dissection Versus No Axillary Dissection in Women with Invasive Breast Cancer and Sentinel Node Metastasis: A Randomized Clinical Trial." *JAMA* 305, no. 6 (2011): 569–75.

Hsueh, Eddy C., and Armando E. Giuliano. "Sentinel Lymph Node Technique for Staging of Breast Cancer." *Oncologist* 3, no. 3 (1995): 165–70.

Imber, Gerald. *Genius on the Edge: The Bizarre Double Life of Dr. William Stewart Halsted.* New York: Kaplan Publishing, 2010.

Jagsi, Reshma, et al. "Contralateral Prophylactic Mastectomy Decisions in a Population-Based Sample of Patients with Early-Stage Breast Cancer." *JAMA Surgery* 152, no. 3 (2017): 274–82.

Katz, Steve J., et al. "Patient Involvement in Surgery Treatment Decisions for Breast Cancer." *Journal of Clinical Oncology* 23, no. 24 (2005): 5526–33.

Kuerer, Henry M., et al. "A Clinical Feasibility Trial for Identification of Exceptional Responders in Whom Breast Cancer Surgery Can Be Eliminated Following Neoadjuvant Systemic Therapy." *Annals of Surgery* 267, no. 5 (2018): 946–51.

———. "Optimal Selection of Breast Cancer Patients for Elimination of Surgery Following Neoadjuvant Systemic Therapy." *Annals of Surgery* 268, no. 6 (2018): e61–e62.

Kummerow, Kristy L., et al. "Nationwide Trends in Mastectomy for Early-Stage Breast Cancer." *JAMA Surgery* 150, no. 1 (2015): 9–16.

Lerner, Barron H. *The Breast Cancer Wars: Hope, Fear, and the Pursuit of a Cure in Twentieth-Century America.* New York: Oxford University Press, 2001.

"Letter from Frances Burney to Her Sister Esther about Her Mastectomy without Anaesthetic, 1812." British Library: www.bl.uk/collections-items.

Morrow, Monica. "Decision Making in Local Therapy for Breast Cancer." *Breast Cancer Research* 9, no. 2 (2007): S8.

Morrow, Monica, et al. "Society of Surgical Oncology–American Society for Radiation Oncology–American Society of Clinical Oncology Consensus Guidelines on Margins for Breast-Conserving Surgery with Whole-Breast Irradiation in Ductal Carcinoma in Situ." *Journal of Clinical Oncology* 34, no. 33, November 20, 2016: 4040–46.

———. "Trends in Reoperation After Initial Lumpectomy for Breast Cancer Addressing Overtreatment in Surgical Management." *JAMA Oncology* 3, no. 10 (2017): 1352–57.

Mukherjee, Siddhartha. *The Emperor of All Maladies: A Biography of Cancer.* New York: Scribner, 2010.

Nash, Rebecca, et al. "State Variation in the Receipt of a Contralateral Prophylactic Mastectomy Among Women Who Received a Diagnosis of Invasive Unilateral Early-Stage Breast Cancer in the United States, 2004–2012." *JAMA Surgery* 152, no. 7 (2017): 648–57.

NSABP Foundation. "Accomplishments of the NSABP in Breast Cancer." nsabp.org/NSABP-Research/Industry-Supported-Clinical-Trials/Accomplishments-in-Breast-Cancer. Accessed March 2018.

———. "For Nearly 50 Years the NSABP Has Been Leading the Way in Breast and Colorectal Cancer Research." nsabp.pitt.edu/NSABP_Timeline.pdf. Accessed April 2019.

Nunn, Daniel B. "Dr. Halsted's Addiction." *Johns Hopkins Advanced Studies in Medicine* 6, no. 3 (2006): 106–8.

Osborne, M. P., et al. "William Stewart Halsted: His Life and Contributions to Surgery." *Lancet Oncology* 8, no. 3 (2007): 256–65.

Oslon, James. *Bathsheba's Breast: Women, Cancer, and History.* Baltimore: Johns Hopkins University Press, 2002.

Paget, Stephen. "The Distribution of Secondary Growths in Cancer of the Breast." *Lancet* 133, no. 3421 (1889): 571–73.

Pimlott, Ken. *2015 Wildfire Activity Statistics.* Sacramento: California Department of Forestry and Fire Protection, 2015.

Shaverdian, Narek, et al. "The Patient's Perspective on Breast Radiotherapy: Initial Fears and Expectations Versus Reality." *Cancer* 124, no. 8 (2018): 1673–81.

Spring, L. M., et al. "Pathological Complete Response After Neoadjuvant Chemotherapy and Impact on Breast Cancer Recurrence and Mortality, Stratified by Breast Cancer Subtypes and Adjuvant Chemotherapy Usage: Individual Patient-Level Meta-Analyses of Over 27,000 Patients." San Antonio Breast Cancer Symposium, December 5, 2018, San Antonio, Texas.

Tadros, A. B., et al. "Identification of Patients with Documented Pathologic Complete Response in the Breast After Neoadjuvant Chemotherapy for Omission of Axillary Surgery." *JAMA Surgery* 152, no. 7 (2017): 665–70.

van la Parra, Raquel F. D., and Henry M. Kuerer. "Selective Elimination of Breast Cancer Surgery in Exceptional Responders: Historical Perspective and Current Trials." *Breast Cancer Research* 18, no. 28 (2016).

van la Parra, Raquel F. D., et al. "Baseline Factors Predicting a Response to Neoadjuvant Chemotherapy with Implications for Non-Surgical Management of Triple-Negative Breast Cancer." *British Journal of Surgery* 105, no. 5 (2018): 535–43.

Yan, M., et al. "Axillary Management in Breast Cancer Patients: A Comprehensive Review of the Key Trials." *Clinical Breast Cancer* 18, no. 6 (2018): e1251–e1259.

Chapter 10. Whole Again

Albornoz, C. R., et al. "A Nationwide Analysis of the Relationship Between Hospital Volume and Outcome for Autologous Breast Reconstruction." *Plastic and Reconstructive Surgery* 132, no. 2 (2013): 192e–200e.

American Society of Plastic Surgeons. *2017 Plastic Surgery Statistics Report*. Arlington Heights, IL. American Board of Plastic Surgery, 2017.

Bennett, Katelyn G., et al. "Comparison of 2-Year Complication Rates Among Common Techniques for Postmastectomy Breast Reconstruction" *JAMA Surgery* 153, no. 10 (2018): 901–8.

Centers for Medicare and Medicaid Services. "Women's Health and Cancer Rights Act." Centers for Medicare and Medicaid Services. cms.gov/cciio/programs-and-initiatives/other-insurance-protections/whcra_factsheet.html. Accessed February 2018.

Champaneria, Manish C. "A Complete History of Breast Reconstruction." In *Breast Reconstruction,* edited by Melvin A. Shiffman, 3–39. New York: Springer International Publishing, 2016.

Fu, Rose, et al. "Abstract 31: The Impact of the 2010 NY State Breast Cancer Provider Discussion Law on Breast Reconstruction Rates: An Analysis of 42,137 Patients from the NY SPARCS Database." *Plastic and Reconstructive Surgery Global Open* 5, no. 4 (2017): 25.

Hartocollis, Anemona. "Before Breast Is Removed, a Discussion on Options." *New York Times,* August 18, 2010.

Hawley, S. T., et al. "Social and Clinical Determinants of Contralateral Prophylactic Mastectomy." *JAMA Surgery* 149, no. 6 (2014): 582–89.

Jagsi, R., et al. "Trends and Variation in Use of Breast Reconstruction in Patients with Breast Cancer Undergoing Mastectomy in the United States." *Journal of Clinical Oncology* 32, no. 9 (2014): 919–26.

Jolie, Angelina. "My Medical Choice." *New York Times,* May 14, 2013.

Lee, Gordon K., and Clifford C. Sheckter. "Breast Reconstruction Following Breast Cancer Treatment—2018." *JAMA* 320, no. 12 (2018): 1277–78.

Santosa, Katherine B., et al. "Long-Term Patient-Reported Outcomes in Postmastectomy Breast Reconstruction." *JAMA Surgery* 153, no. 10 (2018): 891–99.

Schmitz, Kathryn H., et al. "Physical Activity and Lymphedema (the PAL Trial): Assessing the Safety of Progressive Strength Training in Breast Cancer Survivors." *Contemporary Clinical Trials* 30, no. 3 (2009): 233–45.

Solomon, M. P., and M. S. Granick. "Alma Dea Morani, MD: A Pioneer in Plastic Surgery." *Annals of Plastic Surgery* 38, no. 4 (1997): 431–36.

Steiner, Claudia A., et al. *Trends in Bilateral and Unilateral Mastectomies in Hospital Inpatient and Ambulatory Settings, 2005–2013*. Rockville, MD: Agency for Healthcare Research and Quality, 2016.

Uroskie, Theodore W., and Lawrence B. Colen. "History of Breast Reconstruction." *Seminars in Plastic Surgery* 18, no. 2 (2004): 65–69.

Wolfe, Lisa. "Six Years Later, a Triumph over Trauma." *New York Times,* August 28, 1985.

Wong, S. M., et al. "Growing Use of Contralateral Prophylactic Mastectomy Despite No Improvement in Long-Term Survival for Invasive Breast Cancer." *Annals of Surgery* 265, no. 3 (2017) 581–89.

Chapter 11. So Meta

Braun, Stephan, et al. "A Pooled Analysis of Bone Marrow Micrometastasis in Breast Cancer." *New England Journal of Medicine* 353, no. 8 (2005) 793–802.

Carlson, Patrick, et al. "Targeting the Perivascular Niche Sensitizes Disseminated Tumour Cells to Chemotherapy." *Nature Cell Biology* 21, no. 2 (2019): 238–50.

Crimp, Douglas. "Before Occupy: How AIDS Activists Seized Control of the FDA in 1988." *Atlantic,* December 6, 2011. theatlantic.com/health/archive/2011/12/before-occupy-how-aids-activists-seized-control-of-the-fda-in-1988/249302/.

Food and Drug Administration. "FDA Approves New Treatment for Certain Advanced or Metastatic Breast Cancers." U.S. Food and Drug Administration, September 28, 2017. fda.gov/newsevents/newsroom/pressannouncements/ucm578071.htm.

Ghezzi, P., et al. "Impact of Follow-Up Testing on Survival and Health-Related Quality of Life in Breast Cancer Patients." *JAMA* 271, no. 20 (1994): 1587–92.

Leary, Warren E. "FDA Pressed to Approve More AIDS Drugs." *New York Times,* October 11, 1988.

Mariotto, Angela B., et al. "Estimation of the Number of Women Living with Metastatic Breast Cancer in the United States." *Cancer Epidemiology Biomarkers & Prevention* 26, no. 6 (June 2017): 809–15.

National Cancer Institute. "Cancer Moonshot Research Initiatives." National Cancer Institute. cancer.gov/research/key-initiatives/moonshot-cancer-initiative/implementation. Accessed March 2019.

Rosselli Del Turco, M., et al. "Intensive Diagnostic Follow-Up Did Not Improve Survival in Breast Cancer." *JAMA* 121, no. 3 (1994): 77.

Schattner, Elaine. "Notes from the 'Die-In,' a Demonstration to Metastatic Breast Cancer." *Forbes*. October 30, 2015.

Sledge, George W., Jr. "Curing Metastatic Breast Cancer." *Journal of Oncology Practice* 12, no. 1 (2016): 6–10.

Zacharakis, Nikolaos, et al. "Immune Recognition of Somatic Mutations Leading to Complete Durable Regression in Metastatic Breast Cancer." *Nature Medicine* 24, no. 6 (2018): 724–30.

Chapter 12. Prognosis

Flatiron Health and the Information Exchange and Data Transformation Program. "The Role of Real-World Evidence in Regulatory Decision Making, 2018 Flatiron Research Summit." Video uploaded by U.S. Food and Drug Administration, February 13, 2019. youtube.com/watch?v=NK2etXwQg3s.

Genentech. "FDA Approves Genentech's Kadcyla for Adjuvant Treatment of People with HER2-Positive Early Breast Cancer with Residual Disease after Neoadjuvant Treatment." Press release, May 3, 2019.

Gould, Stephen Jay. *Bully for Brontosaurus: Reflections in Natural History.* New York: W. W. Norton, 1991.

Harris, Gardiner. "F.D.A. Regulator, Widowed by Cancer, Helps Speed Drug Approval." *New York Times,* June 2, 2016.

Isidore, Chris. "We Spend Billions on Lottery Tickets. Here's Where All That Money Goes." CNN.com, August 24, 2017. money.cnn.com/2017/08/24/news/economy/lottery-spending/index.html.

I-SPY 2 Trial: Neoadjuvant and Personalized Adaptive Novel Agents to Treat Breast Cancer. ClinicalTrials.gov. Last updated October 31, 2018. clinicaltrials.gov/ct2/show/NCT01042379.

Jain, S. Lochlann. *Malignant: How Cancer Becomes Us.* Berkeley: University of California Press, 2013.

National Breast Cancer Coalition. *The Breast Cancer Deadline 2020.* National Breast Cancer Coalition. breastcancerdeadline2020.org/about-the-deadline/. Accessed June 2018.

Quantum Leap Healthcare Collaborative. "The I-SPY Trials." ispytrials.org.

Rugo, Hope S., et al. "Adaptive Randomization of Veliparib-Carboplatin Treatment in Breast Cancer." *New England Journal of Medicine* 375, no. 1 (2016): 23–34.

Susan G. Komen. "Susan G. Komen Sets Bold Goal to Reduce U.S. Breast Cancer Deaths by 50 Percent in 10 Years." Susan G. Komen, September 13, 2016. ww5.komen.org/News/Susan-G--Komen-Sets-Bold-Goal-To-Reduce-U-S--Breast-Cancer-Deaths-By-50-Percent-In-10-Years.html.

Tversky, Amos, and Daniel Kahneman. "The Framing of Decisions and the Psychology of Choice." *Science* 211, no. 4481 (1981): 453–58.

index

about the author

Kate Pickert is an assistant professor of journalism at Loyola Marymount University, where she teaches investigative reporting, political journalism and other courses. Before joining the faculty at LMU in 2015, Pickert was an award-winning staff writer for *Time* magazine. At *Time,* Pickert covered the debate over and the implementation of the Affordable Care Act and wrote stories about politics, abortion, California and trends in modern American life. She previously worked for *New York* magazine and has written for a variety of other magazines and newspapers. Pickert is a graduate of Columbia University's Graduate School of Journalism and lives in Los Angeles with her husband and daughter.